PROPAGANDA AND INFORMATION WARFARE IN THE TWENTY-FIRST CENTURY

Propaganda and Information Warfare in the Twenty-First Century is the first book to analyze how the technology to alter images and rapidly distribute them can be used for propaganda and deception operations. Such images have already appeared, including altered images of British troops abusing prisoners in Iraq and altered photographs of a 1970s anti-Vietnam War rally with Jane Fonda sitting near 2004 presidential candidate John Kerry.

Using examples from history, Scot Macdonald outlines the principles of propaganda and deception, and presents a history of the use of altered images (both still and moving) in politics, diplomacy, espionage and war. Discussion of the key elements of propaganda, deception and photography serve as the foundation for an examination of why the United States and the West are likely targets for attacks using altered images, how adversaries might use such images against the West and what the possible defenses are against them.

Propaganda and Information Warfare in the Twenty-First Century will be of great interest to students of information war, propaganda, public diplomacy and security studies in general.

Scot Macdonald is Adjunct Professor at the School of International Relations, University of Southern California. He is the author of *Rolling the Iron Dice: Historical Analogies, Regional Contingencies and Anglo-American Decisions to Use Military Force* (Westport: Greenwood Press, 2000).

CONTEMPORARY SECURITY STUDIES

NATO'S SECRET ARMIES
Operation Gladio and terrorism in Western Europe
Daniele Ganser

THE US, NATO AND MILITARY BURDEN-SHARING
Peter Kent Forster and Stephen J. Cimbala

RUSSIAN GOVERNANCE IN THE TWENTY-FIRST CENTURY
Geo-strategy, geopolitics and new governance
Irina Isakova

THE FOREIGN OFFICE AND FINLAND 1938–1940
Diplomatic sideshow
Craig Gerrard

RETHINKING THE NATURE OF WAR
Isabelle Duyvesteyn and Jan Angstrom (eds)

PERCEPTION AND REALITY IN THE MODERN YUGOSLAV CONFLICT
Myth, falsehood, and deceit 1991–1995
Brendan O'Shea

THE POLITICAL ECONOMY OF PEACEBUILDING IN POST-DAYTON BOSNIA
Tim Donais

THE DISTRACTED EAGLE
The rift between America and old Europe
Peter H. Merkl

THE IRAQ WAR
European perspectives on politics, strategy, and operations
Jan Hallenberg and Håkan Karlsson (eds)

STRATEGIC CONTEST
Weapons proliferation and war in the Greater Middle East
Richard L. Russell

PROPAGANDA, THE PRESS AND CONFLICT
The Gulf War and Kosovo
David R. Willcox

MISSILE DEFENCE
International, regional and national implications
Bertel Heurlin and Sten Rynning (eds)

GLOBALISING JUSTICE FOR MASS ATROCITIES
A revolution in accountability
Chandra Lekha Sriram

ETHNIC CONFLICT AND TERRORISM
The origins and dynamics of civil wars
Joseph L. Soeters

GLOBALISATION AND THE FUTURE OF TERRORISM
Patterns and predictions
Brynjar Lia

NUCLEAR WEAPONS AND STRATEGY
The evolution of American nuclear policy
Stephen J. Cimbala

NASSER AND THE MISSILE AGE IN THE MIDDLE EAST
Owen L. Sirrs

WAR AS RISK MANAGEMENT
Strategy and conflict in an age of globalised risks
Yee-Kuang Heng

MILITARY NANOTECHNOLOGY
Potential applications and preventive arms control
Jurgen Altmann

NATO AND WEAPONS OF MASS DESTRUCTION
Regional alliance, global threats
Eric R. Terzuolo

EUROPEANISATION OF NATIONAL SECURITY IDENTITY
The EU and the changing security identities of the Nordic states
Pernille Rieker

INTERNATIONAL CONFLICT PREVENTION AND PEACE-BUILDING
Sustaining the peace in post conflict societies
T. David Mason and James D. Meernik (eds)

CONTROLLING THE WEAPONS OF WAR
Politics, persuasion, and the prohibition of inhumanity
Brian Rappert

CHANGING TRANSATLANTIC SECURITY RELATIONS
Do the US, the EU and Russia form a new strategic triangle?
Jan Hallenberg and Håkan Karlsson (eds)

THEORETICAL ROOTS OF US FOREIGN POLICY
Machiavelli and American unilateralism
Thomas M. Kane

CORPORATE SOLDIERS AND INTERNATIONAL SECURITY
The rise of private military companies
Christopher Kinsey

TRANSFORMING EUROPEAN MILITARIES
Coalition operations and the technology gap
Gordon Adams and Guy Ben-Ari

GLOBALIZATION AND CONFLICT
National security in a "new" strategic era
Robert G. Patman (ed.)

MILITARY FORCES IN 21ST CENTURY PEACE OPERATIONS
No job for a soldier?
James V. Arbuckle

THE POLITICAL ROAD TO WAR WITH IRAQ
Bush, 9/11 and the drive to overthrow Saddam
Nick Ritchie and Paul Rogers

BOSNIAN SECURITY AFTER DAYTON
New perspectives
Michael A. Innes (ed.)

KENNEDY, JOHNSON AND NATO
Britain, America and the dynamics of alliance, 1962–68
Andrew Priest

SMALL ARMS AND SECURITY
New emerging international norms
Denise Garcia

THE UNITED STATES AND EUROPE
Beyond the neo-conservative divide?
John Baylis and Jon Roper (eds)

RUSSIA, NATO AND COOPERATIVE SECURITY
Bridging the gap
Lionel Ponsard

INTERNATIONAL LAW AND INTERNATIONAL RELATIONS
Bridging theory and practice
Tom Biersteker, Peter Spiro, Chandra Lekha Sriram and Veronica Raffo (eds)

DETERRING INTERNATIONAL TERRORISM AND ROGUE STATES
US National Security Policy after 9/11
James H. Lebovic

VIETNAM IN IRAQ
Tactics, lessons, legacies and ghosts
John Dumbrell and David Ryan (eds)

UNDERSTANDING VICTORY AND DEFEAT IN CONTEMPORARY WAR
Jan Angstrom and Isabelle Duyvesteyn (eds)

PROPAGANDA AND INFORMATION WARFARE IN THE TWENTY-FIRST CENTURY
Altered images and deception operations
Scot Macdonald

PROPAGANDA AND INFORMATION WARFARE IN THE TWENTY-FIRST CENTURY

Altered images and deception operations

Scot Macdonald

Routledge
Taylor & Francis Group
LONDON AND NEW YORK

First published 2007
by Routledge
2 Park Square, Milton Park, Abingdon, Oxon OX14 4RN

Simultaneously published in the USA and Canada
by Routledge
270 Madison Ave, New York, NY 10016

Routledge is an imprint of the Taylor & Francis Group, an informa business

Transferred to Digital Printing 2009

Typeset in Times New Roman by
Newgen Imaging Systems (P) Ltd, Chennai, India

British Library Cataloguing in Publication Data
A catalogue record for this book is available
from the British Library

Library of Congress Cataloging in Publication Data
A catalog record for this book has been requested

ISBN10: 0–415–77145–5 (hbk)
ISBN10: 0–415–54500–5 (pbk)
ISBN10: 0–203–96739–9 (ebk)

ISBN13: 978–0–415–77145–0 (hbk)
ISBN13: 978–0–415–54500–6 (pbk)
ISBN13: 978–0–203–96739–3 (ebk)

TO TERRY, BRUCE, PAUL, KRAIG, KEN AND MICHEÁL: MY BAND OF BROTHERS;
TO JENNIFER, SUNYOUNG, DONNA, EMILY, MARLENE AND JANETTE: FOR MAKING MY BROTHERS WHOLE
AND
TO MACDUFF,
WHO NEVER HAD A DEVIOUS BONE IN HIS BODY, AND WAS ALWAYS THE PERFECT IMAGE OF ONE MAN'S BEST FRIEND.
I MISS YOU.

CONTENTS

ACKNOWLEDGMENTS

While the lonely writer in the garret fuels the public's imagination of the writing life, the truth is that writers crave advice, comments and encouragement from the moment an idea for a book strikes all the way through to publication. Col. (Ret) Lloyd J. Matthews, Interim Editor of the US Army War College's *Parameters*, suggested the topic for this book based on an article I submitted. If he had not written a long letter rejecting the article as not being comprehensive enough while suggesting it would make an excellent topic for a book, this work never would have been written.

I would also like to thank Kraig Short, whose clever, and at times devious, mind is only equaled by his dedication to an old friend. His suggestions were insightful and, I hope if we ever go to war, he is on our side to devise deception operations.

My former dissertation advisor, Michael G. Fry, has continued his role as the finest of mentors by providing me with encouragement, advice and a reservoir of knowledge that should have required ten life times to acquire. As he did the first day I met him, he still awes me with his immense knowledge, dry wit and depth of caring for his now former students.

The K.A.M. Institute supported my research with a generous grant, while Ms Allegra and Mr Argyll provided much needed emotional support and encouragement. My parents, Lea and Neil, as they have throughout my life, provided the support only the best of parents can.

Thank you.

INTRODUCTION

In the past, photographs and film were the closest representation of reality available. Propagandists and those seeking to conduct deception operations used crude methods to alter images of real people, events and objects. Such photo-manipulations could usually be detected relatively easily.

Today, those who produce propaganda and practice deception based on images no longer even have to start with reality. Computers allow the creation of any imaginable image, still or moving, with appropriate accompanying audio. The quality of an altered image depends on both the sophistication of the technology used and the skill of the manipulator. Technology will soon provide the ability to alter images perfectly with little or no chance of detection, especially if the time frame for analysis of the image is short, such as during an election campaign or a foreign policy crisis. When that point is reached – if it has not been reached already, the quality of an alteration will depend solely on the skill of the manipulator, and some of them will be exceptionally talented. When combined with the capability of television and the Internet to disseminate images rapidly around the world, the future for the use of altered images for propaganda and deception operations in politics, diplomacy, espionage and warfare is wide open, unstudied and rife with threats.

Throughout history, psychological operations (psyops) and especially propaganda have been powerful weapons. Propaganda was instrumental to the Thirteen Colonies gaining independence and to drawing the United States into Cuba's rebellion against Spain, which ignited the Spanish–American War in 1898. At the close of the First World War, many Germans concluded that British propaganda had contributed significantly to their defeat. At Nuremburg, the Allies charged the Nazis with poisoning the minds of the German people with propaganda, sparking the Second World War and inciting genocide. More recently, in Bosnia and Rwanda during the 1990s, propaganda set off brutal wars of ethnic cleansing, while in 2004 *The 9/11 Commission Report* recognized the power of propaganda when it recommended the use of various types of psychological warfare to improve America's image overseas.

As with propaganda, well-executed deception operations have been powerful tools of generals, intelligence agents and politicians throughout history.

The Mongols deceived their enemies by using stuffed mannequins atop spare horses to inflate the apparent size of their armies. In the US Civil War, General Thomas "Stonewall" Jackson said he planned to "[m]ystify, mislead and surprise" the enemy and the outnumbered Confederacy did so with a vengeance (Latimer 2001: 26–9). In 1941, Adolf Hitler used a deception within a deception to surprise the Soviet Union with an invasion involving three million men, while in 1944, the Allies provided the truth with what Winston Churchill called "a bodyguard of lies" (Brown 1975: 389) to deceive the German high command about the timing, location and size of the Normandy landings. In 1973, the Egyptians employed more than 150 deception operations to achieve surprise when they attacked Israel on Yom Kippur, and in 1991 the Coalition led by the United States used deception to fool Saddam Hussein about the planned location of their attack that resulted in the eviction of Iraqi forces from Kuwait. In 2004, US forces used a ruse to convince insurgents in Fallouja that they would be attacking from the south to clear the Iraqi city when, in fact, they attacked from the north.

These examples illustrate the power of psyops and deception operations to affect the course of nations and the outcome of wars. When they are wedded to the ever-improving modern capability to alter images and rapidly distribute them worldwide, psyops and deception operations will only become more powerful, especially given the dominant role of images in mass communication today and for the foreseeable future.

Whereas in ancient wars propaganda was spread through word of mouth, the advent of movable type some 500 years ago launched a text revolution. Literacy levels increased as the written word became the dominant method of communicating information, supplanting oral-based traditions. By the First World War mass-circulation newspapers and magazines reigned supreme, although newsreels were already hinting at the coming importance of moving images. Motion pictures have worldwide appeal since, unlike text, they do not face any language barriers. By the Second World War, images were paramount, in picture magazines, newsreels and motion pictures, although text and audio, especially Hitler's speeches at political rallies, Churchill's stirring Parliamentary and radio speeches and Franklin D. Roosevelt's fireside chats were still of great importance. Motion pictures, however, heralded the "imaging revolution," "where psychology was as important as logic... where mood rivaled categorical statements, and emotion was as weighty as reason... the end of 'typographic man'" (Hurley 1976: 920) had arrived. The change from words to images was also reflected in the change from the dialog-driven entertainment of the 1930s and 1940s, typified by Vaudeville and such acts as Abbott and Costello, to the image-driven motion pictures of the 1960s until today, which can be watched with the sound muted, yet still convey a story.

The Korean War was the last American war in which written coverage was dominant, and only because film and television cameras were not yet mobile enough to cover the fighting on the rugged peninsula. By the time of the war in Vietnam images had achieved unrivaled dominance. Television had taken over and no great orations like Churchill's "Finest Hour" and Roosevelt's "Day of Infamy" speeches survive in the popular memory. Most people, in fact, remember few, if any words from the Vietnam War; instead, their memories are seared by such images as Buddhist monks committing self-immolation, a naked girl running from a napalmed village, and a Saigon police chief shooting a suspected Vietnam Congress (Viet Cong or VC) prisoner in the street. More recently, the Internet, with its focus on images and graphics, has only solidified images as the dominant form by which most people now receive information.

Even with the increasing dominance of images in mass communications, practitioners of psyops and covert operations have been slow to realize the potential of images, let alone of images that have been altered. Altered images should be defined broadly to include: manipulating images; staging events to be photographed or filmed; changing captions so an image appears to be something other than it really is; focusing on certain things and not others and creating photomontages. All of these forms of image manipulation create an altered reality for the viewer and provide a powerful weapon to those who would practice propaganda or deception.

Although legal, ethical and political constraints limit the ways in which the United States might use altered images in the future, Americans have used propaganda and conducted deception operations in the past and are likely to continue doing so by using some types of altered images. Therefore, such techniques should be studied. The United States also must stay abreast of the threat posed by adversaries' potential use of altered images in order to effectively counter such attacks.

Because the United States is currently the sole superpower, Americans will be fighting conflicts with states and groups that are far weaker militarily for the next decade at least and, if US leaders are prudent, for far longer. Those with the biggest battalions, such as Rome at its height, Napoleon in his later campaigns such as his invasion of Russia, and the United States today have been able to rely more on brute force than on guile to win wars. The belief in using brute force to win was epitomized by British Admiral John de Robeck who remarked, as he watched Turkish machine gunners slaughter Australia–New Zealand Army Corps (ANZAC) troops at Gallipoli in 1915, "Gallant fellows, these soldiers; they always go for the thickest place in the fence" (Ikle 1989: 5). Barring those few commanders who countenance pure attrition warfare, however, the preferred strategy has usually been to win battles and wars by guile, not by military force. Deception and propaganda can secure victory at less cost or, in what Sun Tzu characterized as the acme of martial skill, allow a nation to impose its will on an adversary without even fighting a battle. Although a great power may rely heavily on big battalions and little on guile, invariably, as its strength ebbs or rivals rise, the need for guile, deception and psyops increases in order to avoid defeat in battle or to slow decline in times of peace. Even when a nation is at the zenith of its power, weaker rivals often employ deception and propaganda to level the playing field. This tendency emphasizes the need for powerful nations, such as the United States, to study such methods or suffer sometimes disastrous consequences.

Groups with the capability to conduct operations involving altered images are far from uncommon and will only proliferate as technology and computer expertise spreads. A 1994 US Defense Science Board study found that more than 50 nations have emerging information warfare (IW) capability, while more than 25 states have underground organizations with potentially threatening computer capabilities that include the ability to generate and disseminate altered images (Waltz 1998: 195). Since that study, such capabilities have undoubtedly spread to even more nations and groups.

Unfortunately, the study of the possible use of altered images for propaganda and to support deception operations conducted against the United States has been neglected amid the IW focus on computers and attacks on infrastructure. This bias reflects the common American focus on technology, such as computers, and not on the creative applications of technology. The IW literature sometimes uses the terms "perception management" (Molander *et al.* 1996: xiv) or "media warfare" (Wheatley and Hayes 1996: 3) to suggest that the mass media, including the Internet, can be used to disseminate propaganda or support deception operations. Even when it is mentioned, however, few details are offered and most of the IW literature says little, if anything, about the possible use of altered images in the realms of propaganda and deception operations. This neglect may prove devastating to the United States because even if the United States leads in the development and use of technology to manipulate images, such a lead does nothing to prevent others from using the same technology against the United States.

The use of altered images as propaganda and to support deception operations in politics, diplomacy and war is based on three separate, but overlapping areas: photography, psyops (propaganda) and deception operations. Therefore, this book begins by drawing examples from the history of photography of the five major ways images, whether still or moving, can be altered, and then discusses the media's use of manipulated images, which is becoming far more common. The two major forms of attacks using altered images will be in the realms of propaganda and deception operations, so the middle chapters of the book focus on the principles and past use of psyops, especially propaganda and deception operations. The examples provided, although many of them do not involve images, show how the principles of propaganda and deception are applied, just as they would be applied when such operations included the use of manipulated images. The chapters on propaganda and deception overlap to a degree, just as some propaganda deceives and some deception operations use propaganda techniques.

Researches into the effects of propaganda and deception operations are entire fields in and of themselves. Given that the use of altered images is, historically, relatively new, very little research, however, has focused specifically on the influence of altered images in propaganda and as part of deception operations. This book, therefore, will mention the effects that other researchers have found for specific cases of propaganda and deception, especially those involving altered images. It will not, however, extensively analyze the varying and often debated effects of propaganda and deception in general. Future research, especially when there are more cases involving the use of manipulated images in politics, diplomacy and warfare should focus on analyzing the specific effects of altered images when they are used as propaganda and to deceive.

The chapters on propaganda and deception form a foundation for the final three chapters. The ninth chapter argues that, although the United States is technologically advanced and has many experts in the fields related to the use of manipulated images for propaganda and to deceive, the United States is both an easy and likely target of attacks based on altered images. Chapters 10 and 11 focus on how adversaries might use altered images to support propaganda campaigns and deception operations against the United States, as well as possible defenses and countermeasures against such attacks. Although the United States has the technology and expertise to use propaganda and to conduct deception operations, it often does not take advantage of these techniques. Therefore, the final chapter also discusses how the United States might use such techniques more in the future. The use of altered images for psyops, propaganda and deception in politics, diplomacy and warfare must be studied because, if the pen is already mightier than the sword, the altered image might prove mightier than both.

1

THE LYING EYE

Photography, propaganda and deception

In 1827, the French scientist Nicephore Niepce produced the first photograph. In 1839, William Henry Fox Talbot and Louis-Jacques-Mandé Daguerre announced that they had independently discovered two different methods to capture and fix an image. After further development, the French government bought the rights to Daguerre's process, called the Daguerreotype. Photographers started "faking" photographs from the earliest days of photography. People could not keep their eyes open for the long exposure required by a Daguerreotype, so photographers scratched out the pupils of eyes on the plate or emulsion to make the subject's eyes appear to be open. Further deceptions soon followed. While Daguerre became a celebrity, Hippolyte Barnard, who had helped develop the photographic process, did not. In 1840, Barnard decided that since he was not getting the public attention he felt he deserved for helping to develop photography, he would get attention by other means. He posed himself in a photograph as a drowned corpse and wrote a "suicide" note that said, "The government, which gave M. Daguerre so much, said it could do nothing for M. Bayard [*sic*] at all, and the wretch drowned himself." Although the photograph brought Barnard much more attention, Daguerre still overshadowed him as the inventor of photography (Brugioni 1999: 26).

During the 1840s, rapidly improving methods caused photography to explode in popularity and by 1850, for example, there were 77 photographic galleries in New York City alone. Even so, during the US Civil War newspapers lacked the equipment to transform photographs into half-tone blocks to use in presses to mass-produce the images. Therefore, artists, not photographers, provided illustrations of the war.

It was not until 1900 that the use of photographs in the mass media dramatically increased as American newspapers sought increased readership. In cities with large segments of uneducated natives and non-English speaking immigrants, images, not text, were critical to increased circulation. Each newspaper employed retouchers who altered images for the greatest dramatic effect. Many journalists believed that photo manipulation was normal in part because just developing and printing a photograph involved manipulating contrast, brightness and size. If there was no photograph of an event, one was often faked.

By 1900 faking news events already had a long history. Photographers Henry Negretti and Joseph Zambra, for example, lacked a photograph of British balloonists James Glaisher and Henry T. Coxwell's 5 September 1862 record-setting ascent in a hot air balloon to 37,000 feet. The photographers solved their problem by superimposing the image of the balloonists in their basket against an appropriate background. They then painted in the superstructure of the balloon (Brugioni 1999: 30).

Still photography soon led to the development of motion pictures. The two Lumière brothers, industrialists and inventors from Lyon, developed a system for projecting photographic film onto a screen. On 28 December 1895, the first public showing of a film to a paying audience took place at the Grand Café in Paris. It consisted of 10 short films of about 50 seconds each, including offerings such as "Coming out of the Lumière factories." Within a few years, even battles were being filmed. During the Second Boer War (1899–1902), the British government authorized the first motion picture coverage of a war and just over a decade later the First World War saw the extensive use of photography and newsreels, especially by propagandists.

The inter-war years witnessed a boom in what Frank Luther Mott, later dean of the University of Missouri School of Journalism, in 1924 would call photojournalism. Photojournalism flourished in such magazines as *Life*, *Look* and *Colliers*. By the Second World War, photographs and films were crucial propaganda weapons, with photographers and cameramen present on every front and in every theater. By late 1944, Allied censors in Paris reviewed 35,000 photographs and 100,000 feet of newsreel every week (Knightley 1975: 315).

Television was developed in the 1920s, but did not spread widely until after the Second World War. Contrary to the common belief in the United States, Vietnam was not the first war to be extensively televised. The French–Algerian War (1955–62) was the first covered by television. Since then, television cameramen have become instrumental in reporting politics, foreign affairs and military conflicts around the world.

The images that have become instrumental to reporting on events around the world are powerful. They can frame events or issues, creating reference points for the public. For example, the constant replay of newsreel images of German tanks rumbling across farm fields and Stuka dive bombers hurtling out of the sky have reinforced the commonly held belief that the German military was highly mechanized during the Second World War. This belief, however, is a myth. During the 1939 invasion of Poland, 90 percent of the German army was composed of foot soldiers who relied on horse transport (Deighton 1980: 99). Four years later in 1943, at the height of German mechanization, the typical Wehrmacht infantry division had 5,375 horses and 1,133 horse-drawn vehicles, but only 942 motor vehicles (Deighton 1980: 175). Images still frame events today. The Northern Ireland peace process hit an obstacle in December 2004 when the Irish Republican Army (IRA) refused to be photographed turning in their weapons. The IRA said they had agreed to a "decommissioning" that would put their weapons "beyond use." Photographing the process, the IRA argued, would make it appear that they were surrendering (Daniszewski 2004: A20).

By their ability to frame events, images can help win or lose wars. In a case similar to President George W. Bush's ban on images of military coffins arriving back in the United States from the 2003 intervention in Iraq, Franklin D. Roosevelt felt that images of dead Americans during the Second World War would be so disturbing to the public that he banned their publication until 1943, when he changed his mind and concluded that the public should be shown the sacrifices being made to achieve victory. More than 20 years later images on television, many feel, hammered home to the American public the conclusion that they were losing the Vietnam War. This perception contributed significantly to US withdrawal from the conflict.

The Vietnam War was far from the only conflict in which images have played a crucial role. The Nigerian–Biafran War (1967–70), like Vietnam, was a war of images fought in the court of public opinion, but Nigerian officials never fully grasped the importance of this coverage. "When (to demonstrate their quick justice) they executed before the cameras a Nigerian officer who had a short time earlier (to show his serious intent) executed before the camera a captured Biafran, all they accomplished for most viewers was the piling of one revulsion upon another" (Davis 1976: 698). Biafra, by contrast, quickly learned to get maximum value from photographers and cameramen, who were allowed to roam freely in the territory Biafra controlled. Pictures of starving Biafran children were credible, eliciting global support. A British ITN broadcast, for example, sparked a large Oxfam relief campaign for Biafra. The Nigerian government was far less effective at using television. Images of smiling Nigerian soldiers feeding and playing with children in recaptured areas looked staged, no matter how genuine they sometimes were in fact. In the image war, Biafra soundly defeated their Nigerian adversaries.

More recently, the 1992–3 Somalia operation "lived and died by the television camera" (Adams 1998: 60). Although some in the US government counseled withdrawal even before the bloody fighting of 3 October 1993, the horrifying images of the aftermath of that day dramatically changed US policy. The 3 October mission was a tactical success: many of warlord Mohammed Aidid's senior leadership were captured. Furthermore, 312 Somalis were killed and 814 wounded at the cost of 18 US dead and 84 wounded. A further seven Malaysians were wounded and one killed in a UN force that attempted to rescue the besieged Rangers. By any military calculation, the kill-ratio of 75 to 1 was a massacre.

In the world of images, however, the mission was a US disaster. The global media repeatedly replayed images of Michael Durant, a captured helicopter pilot, who was wounded and had been beaten. The Somalis also mutilated the bodies of five dead US soldiers, dragging one of the corpses through the streets in front of a CNN camera. President William J. Clinton is said to have wept as he watched the images. "By Saturday and Sunday we had won the war," one US military officer recalled, "but on Monday Aidid mounted a strategic attack in the information domain…when bodies started appearing on TV screens, Americans said, 'Wait a minute, nobody told us this was going to happen' – Aidid won" (Adams 1998: 72). Within days, Clinton ordered US troops out of Somalia and cancelled the hunt for Aidid. As in Vietnam, the United States won on the battlefield, but lost the image war.

Besides helping to win or lose wars, photographs can also discredit governments. In April 1984, Alex Libak, a photographer for the Israeli newspaper, *Hadashot*, took pictures of Shin Bet (internal security) officers taking a live Arab hijacker into custody after security officers stormed a hijacked bus. The government later claimed that all of the hijackers had been killed during the fighting. Libak's photographs of the hijacker, who later turned up dead, alive and in the hands of Shin Bet led to a major government scandal (Raviv and Melman 1990: 281–2).

In 1986 images made a crucial difference in proving the validity of a story. Britain's *Sunday Times* only believed and published Israeli nuclear power plant worker Mordecai Vanunu's allegations about Israel's nuclear program when he produced 60 photographs to support his story (Raviv and Melman 1990: 360–78). Images also help credibility in the realm of espionage. Israeli agents often require photographs to support reports from Arab agents working for them, since the Israelis believe that Arabs "tend to exaggerate and often fail to report accurate details" (Raviv and Melman 1990: 427).

More recently, the deaths of two Afghans held at Guantanamo Bay, Cuba, that were ruled homicides, failed to spark much public interest about possible abuses by US troops guarding detainees in the war on terrorism. For two years rumors about ill-treated prisoners circulated, but did not gain national prominence until 2004 when images from Abu Ghraib prison in Iraq exploded into the public's consciousness. As one editorial asked, "Why was [the] pattern of abuse ignored for so long?" (USA Today 2004: 11A). Because there were no pictures: images made the story real.

> In the modern way of knowing, there have to be images for something to become "real."...For a war, an atrocity, a pandemic...to become a subject of large concern, it has to reach people through the various systems (from television and the Internet to newspapers and magazines) that diffuse photographic images to millions.
>
> (Sontag 2003: R16)

Cases ranging from Vietnam and Somalia to the Iraqi prison abuse scandal strongly support Sontag's argument. As an Israeli Minister of Information wrote, "Without television, you cannot have a war" (Hiebert 1995: 335).

A picture is worth a thousand lies: photographic deceptions

> It was true that there was no such person as Comrade Oglivy, but a few lines of print and a couple of faked photographs would soon bring him into existence....Comrade Oglivy, who had never existed in the present,

> now existed in the past, and when once the act of forgery was forgotten, he would exist just as authentically, and upon the same evidence, as Charlemagne or Julius Caesar.
>
> Orwell, *1984*

The word "photography" is derived from Greek words for light and writing, and the "writing" in a photograph is as easy to alter or forge as the written word. There are five main types of photographic deception. Photographers can focus on certain things and not others or they can stage a photograph. They also can alter a photograph after it is taken or change the caption, so that the reader will misinterpret the image. Artists can also create photomontages by combining multiple photographs to create one image.

Focusing on certain things and not others

The media is a searchlight. It focuses its beam on a single subject, then quickly swings around to illuminate something new. Unfortunately, for as much as the media illuminates, even more is lost in the vast region of darkness around the circle of light. Just like the media in general, photographers can use their cameras to capture certain events, while ignoring others and, by doing so, present a distorted picture of reality. Even if "the camera does not lie directly, it can lie brilliantly by omission" (Knightley 1975: 15).

In 1885, the British government sent Roger Fenton, a founder of the Royal Photographic Society and photographer of the royal family, to the Crimean War to attempt to counter negative press coverage. Fenton arrived in Balaclava on 8 March 1855 and started taking photographs. His pictures portray a clean, ship-shape war, bordering on a picnic. There are no wounded, let alone any corpses. The photographs show only happy troops and elegant officers. The pictures were accurate in the sense of showing people and events as they were without for the most part being staged. They were, however, clearly a deception for propaganda purposes that was achieved by focusing on certain things while ignoring others. Such selective depiction of the war was not entirely due to British government pressure. Other photographers not employed by the British government, including James Robertson and the Frenchman Charles Langlois, also failed to portray the horror and carnage of the war. Whatever the reason, their photographs led to a distortedly rosy view of the costly conflict (Knightley 1975: 15).

Besides distorting public perceptions regarding a war, photographs can be used to deceive an adversary about a tactical situation by focusing on certain things and not on others. During the Cold War, the Soviets published photographs of army pontoon bridges stretched across broad rivers as trains and trucks poured across them. No close-up photographs of the trucks or trains were published because someone might have noticed that they were all empty. If they had been loaded, the bridges would have collapsed (Suvorov 1984: 224–5).

The United States has also used the media spotlight to deceive enemies in wartime by showing certain things and omitting others. During Operation Desert Shield, the US military wanted the Iraqis to believe that the Coalition's ground assault would consist largely of an amphibious landing in Kuwait. Great emphasis was accordingly and very publicly placed on the Fourth Marine Expeditionary Brigade. In November 1990, Vice Admiral Henry Mauz, the Chief of Naval Operations for Central Command, took personal command of an amphibious landing exercise off Oman, code-named Camel Sand. Later the same month, another such exercise, code-named Imminent Thunder, took place. Public affairs officers at the Pentagon and in the Gulf carefully drew the media's attention to the exercises and the Marine brigade aboard ships off the Kuwaiti coast. Journalists were told about how the Marines operated, given access to the ships, and encouraged to write about the vital role amphibious landings could play in a land campaign. The Americans knew the Iraqis were relying on the media for information and wanted to convince them that the attacking forces would come from the east through Kuwait, not from the west across the desert as actually happened. Using the media "was considered entirely legitimate." A Pentagon official said, "We told no lies.... The reporters wanted to believe what they saw and simply did not ask the right questions. More fool them" (Adams 1998: 46).

Television coverage of the 2003 war in Iraq also focused on certain things while ignoring others. In a scene flashed across US television screens, what appeared to be an Iraqi multitude cheered as they pulled down a statue of Saddam Hussein in Baghdad. The Iraqi-filled screen suggested a huge crowd but, during the few seconds when the camera pulled back, it showed that the crowd consisted of about 300 people. Furthermore, many in the crowd were foreign journalists. For a city of nearly five million, it was a tiny demonstration. Perhaps most Iraqis were still too frightened to parade in the streets against their former dictator, but if they were, television did not accurately convey that information (Rosenberg 2003: A16).

Many leaders closely manage their own photo opportunities to focus attention on certain things, while downplaying others. Of 35,000 photographs of Franklin D. Roosevelt published when he was alive, for example, none show him in his wheelchair (Smith 2005: A25). The media cooperated with this omission through a gentlemen's agreement not to show the president's disability.

Staging a photograph or film

Instead of focusing on certain things and omitting others, some photographers deceive by actively staging scenes to photograph. Such staging has a long history. Before cameras could be used outdoors, early photographers often used painted backgrounds to create the illusion of an exterior setting. Since then, scenes have been staged to illustrate political and military events, to serve as propaganda and to deceive intelligence services.

Photographers during the US Civil War produced some of the most striking images of war in history. Many, however, were staged. Timothy O'Sullivan and

Alexander Gardner, who worked for Mathew Brady, moved corpses and weapons to create more dramatic scenes to photograph. Possibly the most famous of the staged photographs is a Gardner image of a rebel sharpshooter taken at the Devil's Den two days after the battle of Gettysburg, called *A Sharpshooter's Last Sleep* or *Home of a Rebel Sharpshooter*. Before taking the photograph, Gardner dragged the body of a Confederate soldier about 30 yards to where he lies in the picture, and turned his head toward the camera, resting it on a knapsack to achieve a photogenic angle. The soldier's rifle is still against a stone wall, but since souvenir hunters routinely removed such equipment soon after a major battle, it was almost certainly Gardner's prop. In addition, the weapon in the photograph was a type rarely used by sharpshooters (Brack 1996: 48; Brugioni 1999: 31–2; Nickell 1994: 60–2).

The staging of scenes continued in other war zones. At the start of the 1898 Spanish–American War, photographs of the result of the event that triggered the war, the USS *Maine* at the bottom of Havana harbor after she blew up, were shot in a fish tank in New Jersey (Rosenberg 1997: F1). During the Italian campaign in Abyssinia in 1935–6, photographers were not allowed at the front. Responding to their editors' demands, many photographers staged photographs. Herbert Matthews, a *New York Times* reporter, later said that 99 percent of the photographs published about Abyssinia were faked. Joe Caneva of the Associated Press showed more initiative than most when he convinced the Italians to maneuver about 50 tanks and several companies of soldiers before his camera. Later his pictures of the Italian maneuvers appeared in newspapers around the world as a tank charge against the Abyssinians (Knightley 1975: 186). After the Six Day War, the Israelis photographed Egyptian soldiers in their underwear next to undamaged Soviet tanks and in other unheroic poses to convey the message that the Egyptians were cowards (Andrew and Gordievsky 1990: 498).

Staging events for the camera did not stop at still photographs. During the 1898 Spanish–American War, newsreel films showed the decisive Battle of Manila Bay. Audiences, however, were not told that the exciting battle scenes had been shot with models in a fish tank. During the Second Boer War, the Edison Company, the first motion picture company, lacked actual footage of the war, so they staged battles to film (Rosenberg 1997: F1). Edison was not the only company that made do with staged reenactments. A British newsreel showed Boers attacking a Red Cross tent full of wounded as British doctors and nurses valiantly defended their helpless patients. If the film had been real, the British would have had reason to worry, since the "Boers" in the film were actually "attacking" across Hampstead Heath, which is in a London suburb (Knightley 1975: 75).

Even rebels staged scenes for motion picture companies. In 1913, Francisco "Pancho" Villa was short on cash and seeking US support for his revolution in Mexico. Realizing the propaganda value of motion pictures, Villa sold the film rights to his battles for $25,000 to the Mutual Corporation. Villa then scheduled executions for the camera and re-staged battles after the fighting had ended, using the actual dead who were still lying on the ground to make the staged battles appear more authentic (Rosenberg 1997: F1). The Mutual Corporation also made Villa more marketable to US audiences by lightening his skin, trimming his mustache and dressing him in the popular American image of a bandito with crossed ammunition belts over his chest and a large sombrero. The resulting film, *The Life of General Villa*, debuted in New York in April 1914 (Kraul 2004: E27).

D.W. Griffith, who had convinced his partner at the Mutual Corp., Harry Aitken, to buy Villa's film rights, also filmed staged First World War battles in the United States. He felt that film of the actual First World War battles did not look realistic enough (Rosenberg 1997: F1). Griffith was far from alone in staging films and photographs during the First World War. A staged Russian film showed "German" nurses, dressed as religious sisters, stabbing wounded Russian soldiers on the battlefield (Ponsonby 1928: 138). Stills also can be taken from motion pictures to represent events, such as the Russians did with a photograph of soldiers apparently shooting into a peaceful crowd at the Winter Palace on Bloody Sunday, 9 January 1905. The image is a still from the 1925 film *The Ninth of January* (King 1997: 186).

Many combat films of the Second World War were clearly staged. Some films show shells exploding near soldiers, yet the men appear unharmed. The rounds were undoubtedly training rounds and the "combat" images are of training exercises. The Nazis, for example, staged combat films and photographs. One of the best German combat magazines was *Signal*. An extremely effective propaganda organ, *Signal* probably used photographs from training exercises and labeled them as combat shots. It also published suspiciously high quality air-to-air combat photographs. For example, *Signal* published an excellent photograph of a Me-109 closing in on a Spitfire with the caption:

> In the cloud covered English skies it is the same story every day. A Messerschmitt has discovered a Spitfire and turns toward it. The hunter becomes the hunted. The RAF pilot banks down and away, trying to escape; but the Messerschmitt stays with him, gains on him and peppers him with machine gun fire until he crashes.
>
> (Brugioni 1999: 168–9)

In fact, the Spitfire had been shot down and then repaired to fly again and play the role of the Me-109's compliant victim. The ruse was exposed because the Germans put various insignia and markings on the Spitfire in the wrong place.

Not all of the Nazi's staged images were nearly so benign. In 1944, building on the success of a Potemkin village created for visiting Red Cross officials, the Nazis produced *The Fuehrer Gives the Jews a City*. Staged for neutral states, the film was designed to show that the 50,000 Jews forced to live in Theresienstadt, a 1790 fort near Prague, were living an idyllic existence. The Nazis forced Kurt Gerron, a decorated Jewish First World War veteran with experience in 27 motion pictures, including Marlene Dietrich's 1930 classic *Blue Angel*, to direct the film. In reality, many of the "actors" were dying of typhus and starvation, and were regularly being shipped to Auschwitz and replaced by other Theresienstadt inmates (Thomas 2004: E13; Turan 2004: E6).

Staged photographs are also used in the clandestine world of spies. After the Second World War, the Soviets conducted a deception operation in which staged photographs played a key role. They used photographs of the aftermath of a staged battle between Soviet tanks and anti-Soviet Polish groups, including WIN, a Polish acronym for freedom and independence, to convince the CIA and MI6 that there was an active Eastern European resistance against the Soviets. The West then supplied WIN with arms and, more importantly, information about anti-Soviet Poles living in Poland. Kim Philby, the English Soviet mole, fed information on the deception operation back to the KGB. The Soviets used this information to refine their operation to ensure that the West believed that the resistance groups were real. For four long years the deception worked, leading to the execution of many Polish anti-Soviets and dozens of Western agents who had parachuted behind the Iron Curtain to aid the bogus anti-Soviet forces. The operation only ended when the Polish Communist government publicized the deception to discredit the CIA and MI6 (Epstein 1989: 34–9).

Egyptian security officers used staged photographs to trick Libyan leader Mu'ammar Qaddafi into believing that his assassins had killed Abdul Hamid Bakoush, a former Libyan Prime Minister and Qaddafi opponent, who was living in Egypt. The Egyptians had captured the assassins before they could complete their mission and then staged a photograph showing Bakoush lying dead in a pool of blood. After Egypt published the photographs, the Libyan news agency proclaimed that Bakoush had been executed because he "sold his conscience to the enemies of the Arab nations and the Libyan people." Then Egyptian President Hosni Mubarak revealed that the photographs had been staged and that Bakoush was alive, much to Qaddafi's embarrassment. If Qaddafi had spent more money on photo-interpreters than on assassins, he might not have been so easily fooled. The photograph showed Bakoush lying on the floor with his head to the side, but with blood flowing down his face over his chin. The blood could not have flowed down his face, except in the unlikely event that he had remained standing after he was fatally shot (Brugioni 1999: 22–3).

The health of foreign leaders is crucial in that it can significantly affect a government's decision-making. When Mao disappeared from public view in late 1965, for example, there was widespread speculation about his health. On 16 July 1966, the Chinese News Agency published a report with a photograph that Mao had swum nine miles downstream in the Yangtze River. Careful analysis showed that the man swimming in the photograph had the same mole on his chin and receding hairline as Mao. After comparing the photograph of the man swimming to photographs that were known to be of Mao, however, the man in the Yangtze photograph was found to be a double. His ears gave him away. Ears are as distinctive as fingerprints and proved that the photographs were of a person other than Mao (Brugioni 1999: 155).

Political events are often staged to be photographed. Staging of presidential events arguably reached a high point during Ronald Reagan's tenure, although

George W. Bush also has an exceptional team of location scouts, technical experts and managers who ensure that he appears presidential at every event. Even apparently freewheeling town hall meetings and talks with military personnel are often "precooked" with certain "everyday" people chosen because of the questions they want to or have been instructed to ask (Chen 2004: A15; Klein 2004b: 21). In 2005, for example, the Bush administration was criticized for a carefully staged videoconference between Bush and 11 military personnel serving in Iraq who provided an upbeat view of the war (Vieth and Mazzetti 2005: A14). The management of such events is so accepted by politicians that they are upset when reality intrudes. In 2004, Secretary of Defense Donald Rumsfeld was clearly unsettled when he was questioned during a soldier-style town hall meeting about troop safety by a National Guardsman in Kuwait on his way to serve in Iraq. Rumsfeld was even more upset when it turned out that a reporter had helped the soldier craft his question about the lack of available body and vehicle armor (Mazzetti 2004: A8). Apparently some politicians believe that only they should stage such events; audiences should not.

If anything, staged images appear to be becoming more common. In 1995 Kashmir rebels attempted to convince the United States that an American hostage they were holding had been shot in the stomach by an Indian Army patrol. The image showed the hostage with bloody bandages around his stomach, but none on his pants. Analysts concluded that it was a fake (Junger 2002: 88). The famous images of a Baghdad crowd celebrating as a US Marine vehicle toppled a statue of Saddam Hussein on 9 April 2003 were presented as a spontaneous outpouring of anti-Hussein feelings. It turned out, however, that a Marine colonel staged the event, which was exploited by an alert US Army psyops team, who arranged for the media to be present (Zuchino 2004: A28). In May 2004, the British government charged that images that showed British abuse of Iraqi detainees had been staged. The government stated that a truck in some of the images had, according to the Ministry of Defense, never been in Iraq (Associated Press 2004b: Online). The *Daily Mirror* later apologized for publishing the altered images, which turned out to have been faked by a British soldier (Associated Press 2004c: Online). In February 2005, a militant group posted an image of a US soldier said to have been kidnapped in Iraq on its website. A company that made large action figures, however, said that the image appeared to be of one of their toys – a large, realistic US soldier toy – that had been sold at US bases in Kuwait. The image did, indeed prove to be a hoax (Morin 2005: A6).

Altering a photograph or film

Even if a photographer does not stage a photograph, the image still can be altered after the photograph is taken by either adding or deleting details. Adding details to photographs began as soon as cameras were used outdoors. The collodion emulsion used by early photographers was overly sensitive to blue light, so the sky appeared in photographs as a white void. Undaunted by the failings of technology, many landscape photographers simply mimicked their artist colleagues and painted a blue sky and clouds onto their landscape photographs (Brugioni 1999: 37). Some early photographers went even further, using double exposures to add what appeared to be ghosts and spirits to photographs. Such ethereal images were popular in the mid- to late-1800s.

During the US Civil War, engravings had to be made of combat-photographs in order to print the images in newspapers. The engravers, however, often enhanced their engravings by adding bodies and debris to make battle scenes more dramatic than the original photographs. Like the engravers, photographers also often added images to pictures they took. This was relatively common in group shots when someone failed to appear for a sitting. Mathew Brady, for example, added General Frank P. Blair to a photograph, *Sherman and His Generals, 1865*, for which Blair was not present (Brugioni 1999: 32). To enhance his reputation, Stalin was often added to historic events at which he had not been present, and people were added to photographs of his speeches to make crowds appear larger. Dignitaries who were not at May Day celebrations in Moscow were also routinely added to the reviewing stand atop Lenin's Mausoleum (King 1997: 78–9). More recently, during the 1992 Bush–Clinton election campaign, an official White House photograph showed three generations of the Bush family at a gathering. Various family members, however, had been added (Brugioni 1999: 49).

Far more than individuals can be added to images. During the 1930s, as war loomed, many states added weapons and troops to pictures to enhance perceptions of their military strength. For example, one US photograph showed 32 Martin bombers over Dayton, Ohio. Four photographs of eight planes each, however, had been combined into one image that appeared to show 32 planes (Brugioni 1999: 51–2). In 1982, Palestinians in Beirut released a series of photographs of an Israeli air raid in Lebanon. One such photograph even appeared on the front page of *The Washington Post*. The aircraft in the picture, however, were mirror images; the image of one plane had merely been turned over and added to the photograph as if two planes, not one, were making bombing runs. Furthermore, the planes could not have been flying as close to the ground and at the angles they appeared to be, unless they both crashed seconds after the photograph was taken (Brugioni 1996: 166–7). In a more recent case, in 2005 a former Florida Secretary of State accused some newspapers of adding make-up to her face "to poke fun at her" (Stacy 2005: A5).

As with adding details, deleting details also began almost as soon as the camera was invented. Before photography, most people had only seen themselves represented by portrait artists, who were inclined to make their subjects, who usually paid their commissions, look as flattering as possible. The camera, however, had no such inclination and people were often shocked by their appearance in Daguerreotypes. To remedy this problem, artists were hired to remove blemishes from early portrait photographs. The practice continues today in everything from graduation and wedding portraits to glamour and fashion photographs, in which computer tools remove pounds from waistlines, and scars and blemishes from necks, hands and faces.

Political leaders, like models and movie stars, also want to look even better than their best, so they often have their photographs retouched. Stalin's face, for example, was pockmarked, so his skin was altered to delete any blemishes and his mustache was darkened (King 1997: 9, 98–9). Even the trademark rose-colored birthmark on Soviet leader Mikhail Gorbachev's head was removed from many of his official photographs. Although sometimes done because of vanity, such alterations also can be used to disguise health conditions that could have political repercussions. Photographs of French President Francois Mitterand were retouched to reduce the swelling of his face and neck caused by cortisone shots for his prostate cancer, while Kim Il-Sung, North Korea's leader for decades, had a large lymphoma on the base of his neck that was always airbrushed out of photographs (Brugioni 1996: 163).

In addition to retouching, politicians sometimes totally remove unwanted people. Obliterating someone for political or traitorous acts has a long history, and can be traced back at least as far as Ancient Rome. Called *damnatio memoriae*, the Roman Senate would order a deposed emperor's face chiseled off sculptures. Stalin continued the Roman practice, albeit mainly in photographs rather than on statuary, although he did have statues and artworks obliterated. Politicians, generals, musicians, writers, artists, athletes, scientists and cosmonauts who had fallen from Stalin's favor were airbrushed out of photographs and often crudely deleted with scalpels or ink from works of art and books. It is unclear why the official deletions were often so crude and obviously done, but it may have been done to serve as a warning to those who even considered opposing Stalin (King 1997: 9, 13).

Leon Trotsky, Lev Kamenev, Nikolai Bukharin and Grigori Zinoviev, to name just a few Bolshevik leaders, were removed from photographs after Stalin ordered their execution. In a 1940 photograph, for example, Nikolai Yezhov is strolling with Stalin along the Moscow-Volga Canal. Yezhov later fell from Stalin's grace and was removed from the photograph. His image was replaced by the waters of the canal, which seemed only fitting, since Yezhov had been Commissar of Water Transport. So many people were deleted over the years from a group photograph of a celebration of the second anniversary of the Communist Revolution in 1919 that by 1967 only Lenin remained of the major revolutionaries (King 1997: 46). Yet another Soviet group photograph lost 11 of the original 41 Bolshevik leaders over the years (King 1997: 54–5). Ironically, after his death in 1954, Stalin's successors deleted him from many official photographs.

The practice of deletion did not stop with Stalin's death. Well into the 1970s, the Soviets continued to delete those who had fallen from the Communist pantheon of heroes and leaders. The practice was also a favorite of all Communist and Fascist regimes. After he resigned as premier of Czechoslovakia on 17 April 1969, Alexander Dubcek was removed from many photographs of past national celebrations and events. Similarly, photographs of Mao's mourning ceremonies in September 1976 were widely distributed, but in the November 1976 *China Pictorial* the Gang of Four, who had been arrested on 12 October, had been deleted from all the photographs (Brugioni 1996: 148).

Even democracies are not immune to deleting unwanted people. During the 1980 US presidential campaign the Democratic National Committee (DNC) prepared a brochure entitled *Small Town America Can Depend on Jimmy Carter*. They decided to use a photograph of Carter signing the Rural Development Act in the brochure. There was only one problem: Republican Congressman John Hammerschmidt was in the photograph. Not wishing to have a Republican in a Democratic brochure, the DNC airbrushed Hammerschmidt out. The deletion backfired when it was noticed and the media publicized the alteration. The Republicans reaped a publicity windfall from the story (Brugioni 1996: 165–6).

Images are commonly manipulated in motion pictures today to entertain, but some films have been altered for political and propaganda purposes. In 1934, Metro-Goldwyn-Mayer (MGM) produced three altered newsreels linking Democratic California gubernatorial candidate, Upton Sinclair, with radicals, immigrants, communists and hobos. Days before the election, the films were shown in movie houses across the state. Sinclair supporters rioted in the theaters, believing that he had betrayed them, and he lost the election (Rasmussen 2003: B4).

After France and the Low Countries fell in 1940, Hitler was shown in a newsreel dancing a jig of joy at the demise of his enemies. The jig had been created by spooling or repeating a series of images of Hitler stamping his foot, which made it appear as if the Nazi leader was dancing a jig (Brooks 2002: 314). Besides making dictators dance, altered films can take an interview and match answers to questions other than those asked of the subject, as was suspected in an East German film of interviews with US prisoners of war (POW) in Hanoi during the Vietnam War (Preston *et al.* 1976: 855–62).

False captioning

While some photographs deceive through staging and alteration, others are of real events. They deceive not through the image, but through false captioning. Such mislabeling is especially common during wartime. For example, in 1917 *The Times* of London printed a horrific story about a German factory that produced glycerin for munitions by melting down the corpses of German soldiers. The macabre story was one of the most powerful of the thousands of atrocity stories that appeared during the First World War. By one account, the story started when British Brigadier-General J.V. Charteris, Chief of Intelligence, happened to have two photographs on his desk. One was of German corpses being hauled back from the front in wagons. The other showed dead horses on their way to a soap factory. Charteris later bragged at a dinner party that he simply switched the captions, changing soap factory to glycerin factory, and then sent the story to Shanghai for release. He thought the photograph would have a more powerful impact in China than in the West because the Chinese revered and worshipped their ancestors. Charteris hoped that the story would then play back to Europe, which it did with long-lasting effects: the story did not die down until about 1925 (Knightley 1975: 105–6; Ponsonby 1928: 102–13).

In another example from the First World War, a photograph by Berliner Karl Delius showed the delivery of mailbags at a field post office in Kavevara in the Balkans. German soldiers are standing around as Serbs unload the bags. The London *Daily Mirror* reproduced the photograph on 3 December 1915, with the caption: "Made to Wash the Huns' Dirty Linen," and "The Blond [*sic*] beasts are sweating the Serbians, who are made to do the washing for the invaders. Like most customers who do not settle their bills, they are full of grumbles and complaints. Here a pile has just arrived from the wash" (Ponsonby 1928: 135–6). A photograph from a German newspaper of Russian and French internees lining up for food distribution in Germany was reprinted in the London *Daily News* in April 1915 with the caption, "German Workers Feel the Pinch." The *Berlin Lokalanzeiger* of 9 June 1914, published a photograph of three cavalry officers holding up cups and other trophies they won at the Army steeplechase in the Grunewald. The photograph, however, was reproduced in British and Russian newspapers as Germans holding aloft loot from their conquests. Still another photograph of German officers inspecting munitions cases was reproduced in the 30 January 1915 British *War Illustrated* as "German officers pillaging chests in a French chateau." The same newspaper reproduced a photograph of a German soldier bending over a fallen comrade, with the caption: "Definite proof of the Hun's abuse of the rules of war. German ghoul actually caught in the act of robbing a Russian." Not to be outdone, the *Illustrated War News* of 29 December 1915 pictured a sergeant holding up a sort of cat-o'-nine tails whip. The caption: "What was it used for? A German Whip Among a Collection of War Trophies.... The presence of the whip is of curious significance." The whip shown was in reality a rather mundane German carpet beater (Ponsonby 1928: 136–8).

Although the British were the most effective at using altered captions during the First World War, every nation involved in the conflict used the same tactic. Photographs of corpses from a 1905 pogrom in Russia were circulated in France under a caption attributing the scene to German atrocities. Another photograph of a crowd taken on 13 July 1914, before the war, at the royal palace in Berlin was reproduced in *Le Monde Illustre* on 21 August 1915 with a caption that said the scene was of a German crowd celebrating the sinking of the *Lusitania*.

False captioning was relatively uncommon during the Second World War, but the Soviets often used such deceptions during the Cold War. For example, the Soviets published a photograph in the late 1950s of what they claimed was one of their "large missile static test stand[s]." The United States was worried, since the photograph appeared to show that the Soviets had made significant advances in rocket design and development. Their worries, however, were unwarranted. After comparing hundreds of photographs, the Soviet photograph was discovered to be of a Rocketdyne test stand in Santa Susana, California, with a new Soviet caption (Brugioni 1996: 68).

The use of false captions has continued in Russia even after the end of the Cold War. The official Russian news agency, ITAR-TASS, for example, released a photograph on 15 July 1995 that was said to be of Boris Yeltsin in his hospital room at Moscow's Central Clinical Hospital, after he was admitted on 11 July for heart trouble. Unfortunately for the Russians, the Associated Press acquired a video containing an identical scene of Yeltsin in the same sport shirt but in front of drapes with a bank of four telephones at his left elbow. Sergei Medvedev, who became Yeltsin's press secretary, shot the video on 2 April 1995 in the southern resort town of Kislovodsk, where Yeltsin was vacationing. When confronted by the apparent alteration of the film and false captioning, Dimitry Ardamatsky, of Yeltsin's press service, asked, "If the President wears the same shirt in both old and new pictures, so what?...Do they suppose he should change his shirt three times a day?" Even so, Russian television soon showed Yeltsin in a new, different sport shirt (Brugioni 1996: 152).

The Middle East also has seen its fair share of falsely captioned images. One photograph bore the caption: "One can see how an old Palestinian woman is begging a Phalangist soldier not to separate her from her children, while the [Beirut] area of quarantine is burning and small children are raising their hands." The uniformed soldier, however, is armed with an American M-1, which the Palestinians used in the late 1950s and early 1960s. Their enemies, the Phalangists, used AK-47s (Brugioni 1996: 167).

The media also has used false captions. During the Russo-Finnish War, for example, an editor wanted newsreel film of the Soviets bombing the Finns. He asked around, cabled correspondents and searched everywhere for film, but could not find a single frame from Finland. Desperate, he used film of a bombing from the Spanish Civil War and hoped no one would notice the absence of snow (Knightley 1975: 226).

As the previous example highlights, even though films do not have captions, their meaning can be altered by their presentation. Sometimes this deception is discovered and the truth prevails. On 22 August 1955, Georges Chassagne, a French cameraman for Fox Movietone covering the war in Algeria, was invited to film some prisoners of war. He arrived and, as he filmed, a French auxiliary shot a prisoner before him. The film was shown in North America, but not in France. Stills from the film then appeared in the October 1955 *Life*, but France banned the issue. On 29 December 1955, however, *L'Express* published the *Life* photographs. The French Ministry of the Interior then accused Chassagne of bribing the policeman to shoot the prisoner so he could film the execution. Chassagne denied the accusation and four days later the government retracted the charge, grudgingly admitting that there had been no bribe and that Chassagne had not instigated the shooting (Knightley 1975: 362).

Misinterpretation of a film played a major role in a Soviet deception operation that was designed to persuade the United States not to pursue an anti-ballistic missile (ABM) system. The Soviets showed footage of a missile hitting another missile. US analysts were shocked because the film appeared to show that the Soviets had developed an effective ABM. If the film had been taken during a real test, it would have been cause for alarm, because the United States is still trying to develop an effective ABM at the start of the twenty-first century. The Soviets, however, had filmed an interception between an ABM and a missile for which the trajectory was known. Such an interception is easy: the challenge is to hit an adversary's missile for which the trajectory is unknown. As with all good deception operations, the film was not used by itself. The Soviets supported the film ruse by displaying a new ABM at parades and by constructing a building near Moscow that emitted strange electronic signals. The Soviets then leaked that this was a new ABM guidance station. To further reinforce the deception, Soviet agents in the West were ordered to show no interest in gathering any information on US ABM systems. By no longer seeking Western ABM information, the Soviets hoped that the Americans would conclude that the Soviets were so far ahead in developing ABM technology that they felt they could learn nothing new from the West. The result of the well-coordinated deception? The United States negotiated SALT I, which severely limited ABM system deployment. Although other factors contributed to the treaty, the deceptive Soviet missile film played a crucial role in coloring the views of US policy makers (Suvorov 1984: 164–6).

Photomontages

A photomontage is created by combining two or more photographs or other separate pieces of graphic material into one work. In 1857, Oscar Rejlander created the first major artistic photomontage by combining 30 negatives of separate figures and groups to create an allegorical composition, *The Two Ways of Life*, about two men setting off in life on different paths (Brugioni 1996: 28–30). Photomontages quickly became common. Because so few authentic photographs of Abraham Lincoln existed, for example, photographers met the great demand for portraits after his assassination by creating numerous photomontages in which Lincoln's head was put on the bodies of other notables, including Alexander Hamilton, Martin Van Buren and even Southern secessionist statesman John C. Calhoun (Brugioni 1996: 33). Photomontages were popular in the First World War when soldiers were often photographed smoking cigarettes with far-away, wistful looks in their eyes. The images of loved ones were then added to the clouds of smoke (Brugioni 1996: 45).

Although most photomontages were and are created for artistic reasons and are not intended to deceive, they can be powerful weapons in the propagandist's arsenal. An early example of the propagandistic use of photomontages occurred in Berlin after the First World War, when John Heartfield, an artist and founder of the German Communist Party, used photomontages to criticize Hitler and the Nazis. In 1932, for example, he created a photomontage of Hitler swallowing gold coins, which accumulated in his stomach, with the caption: "Adolf, the Superman: Swallows Gold and Spouts Junk" (Brugioni 1996: 47–8).

During the 1930s, photomontages created for propagandistic purposes flourished in magazines and books, and on posters and billboards. The Soviets used photomontages to celebrate Lenin's electrification program, while Benito Mussolini used them to promote his Fascist policies. The United States also used photomontages for propaganda. The largest photomontage ever made, at about 100 feet high and 118 feet wide, appeared in New York's Grand Central Station in 1941, and was designed to promote the sale of defense bonds by showcasing US military might (Brugioni 1996: 48–9, 52–3).

During the Cold War, both East and West Germany used photomontages extensively. Dr Konrad Adenauer became chancellor of West Germany in 1949 and supported a strong NATO. The East German press widely circulated a photomontage depicting Adenauer in a Nazi uniform, suggesting that he was the heir to the Third Reich. The West Germans retaliated by creating a photomontage of Walter Ulbricht, the East German Communist leader dressed in a Soviet private's uniform and claimed this was the highest military rank Ulbricht had attained with the Soviets. The West German photomontage must have stung because soon after its appearance the photomontages of Adenauer abruptly disappeared (Brugioni 1996: 168).

The North Vietnamese, who also used photomontages extensively, monitored the US media and found that the American public reacted strongly to stories about the massacre of civilians. On one occasion, the North Vietnamese combined a *Newsweek* cover photograph of General William Westmoreland, US commander in Vietnam, with a massacre scene and circulated it as proof that the US military was deliberately killing civilians (Brugioni 1996: 145).

Photomontages also have been common fare in political campaigns. Such cases often illustrate the importance of timing when using altered images for propaganda and to deceive. After Senator Joseph McCarthy charged that there was widespread Communist infiltration of the State Department, a Senate subcommittee led by Millard Tydings, a Democratic Senator from Maryland, investigated and reported that the charges were false. In the 1950 Maryland senatorial campaign, incumbent Tydings ran against Republican John Butler, who was supported by McCarthy. A few days before the election, Butler supporters widely distributed a four-page tabloid, *For the Record* that showed a picture of Tydings conversing with Earl Browder, a US Communist leader, in a poorly done photomontage. Browder's head was too large and light fell on Tydings' head from one direction and on Browder's head from another. Although the fakery was quickly detected and exposed, Tydings blamed the damage done by the photomontage for his loss in the election. It was later determined that the photomontage was created by a photo-technician at the *Washington Times-Herald*, a newspaper that was controlled by a political opponent of Tydings. Butler's campaign manager pleaded guilty to violating Maryland's election laws and paid a $5,000 fine, but Butler never apologized. He said the tabloid images were "merely designed to illustrate the positive attitude displayed by Tydings toward Browder" (Brugioni 1996: 127–9). Also in 1950, a similar incident occurred when Congresswoman Helen Douglas lost a US Senate race to Richard Nixon after photographs of her embracing a well-known Communist, shaking hands and beaming at another Communist, and kissing a black person were published. All the pictures were photomontages and Douglas, like Tydings, blamed the deceptive images for her defeat (Brugioni 1996: 129).

Photomontages have become even more popular today as rapidly evolving technology makes them increasingly easy to create and difficult to detect. Satirical photomontages, many lampooning politicians, such as President Bush and his father fishing in New Orleans amid the devastation of Hurricane Katrina in 2005, are common on the Internet. Other photomontages, however, are more serious and deceptive or propagandistic. During the 2004 Democratic primary, a photograph of front-runner John Kerry was published that showed him at an anti-Vietnam War rally in 1970 seated near Jane Fonda. The picture was, in fact, a photomontage to which Fonda had been added (Rutten 2004a: E14). After Israeli leader Yitzhak Rabin was assassinated in 1995 there was acrimonious debate in Israel about the use of photomontages for propaganda purposes. Before his assassination, the opposition party, Likud, had called Rabin a traitor and murderer. Leaflets were distributed that showed him with hands dripping with the blood of Jews killed by Hamas, the militant Islamic group, and one photomontage even showed Rabin in an SS uniform, weeks before he was killed (Brugioni 1996: 197).

Benefits

Although altered images are often used to deceive, it would be a glaring omission not to mention some of the many beneficial ways in which altered images serve us. Far from being deceptive or causing harm, they can be used for great good. Many photographs from the space program, for example, are enhanced to show celestial objects more clearly. The ability to alter images also is used by police to help reconstruct facial features from skulls for criminal investigations, including missing person and child abduction cases. In the later case, age-progression techniques are used. Similar methods, called forensic anthropology, helped identify the skeletons of Tsar Nicholas and his family, as well as many of the "disappeared" in Argentina after the 1976–83 Dirty War and in the Balkans after the fighting in the 1990s. Apart from identifying victims of crime, war and political repression, the ability to alter photographs can help witnesses describe a suspect using computer-assisted facial reproduction. Altered images are also used extensively in medicine by using color to enhance certain structures, organs or tumors in the body. Some benign uses of altered images are purely for entertainment, such as morphing, a contraction of metamorphosis, which uses computers to make it appear as if one object or person smoothly transforms into another. As just one early example, this process was used in the motion picture *Terminator 2: Judgment Day* (1991), in which a female terminator morphs into a metallic state and then into a man, in the form of actor Robert Patrick (Brugioni 1996: 173–91). All such uses of altered images benefit us and allow us to see things that we could not otherwise see.

2

ALTERED IMAGES

Here, there and everywhere

The technology already exists to alter still and moving images with appropriate accompanying audio in almost any imaginable way. Current technology also allows the creation of entirely new computer generated images (CGI) of events that have never occurred. Such technology could easily be used to create propaganda and provide images to support deception operations, especially targeted against the world's most image-based society: the United States. Other image-saturated societies from Britain to Japan, Australia to Israel are also vulnerable to such attacks.

Digital technology allows editors of fashion and entertainment magazines, using Adobe Photoshop and other programs, to airbrush out wrinkles, freckles and blemishes, as well as much more, from photographs of models and celebrities. Photoshop even has a "healing tool" that removes inconsistencies from skin with just one click. Cut-and-paste functions have also become so advanced that when two images are merged into one, they appear seamless. Computers now can be used to alter colors, merge photographs, add, delete, crop, etch, strip and do everything that could be done with older photographic techniques. Digital alterations are far easier and faster to make, and are much harder to detect because computers leave few, if any indication that the image has been manipulated. Digital alterations are also reversible. Manipulators can change an image, then change it back and try again until they are satisfied that the alteration is undetectable. Furthermore, the disk with the original image on it can be erased, leaving no record of the original image. With no negative, it is much more difficult to detect a fake. The difference digital-based images make over film-based photography is easy to comprehend if you compare trying to alter something that has been written on a typewriter (analogous to film) to something written on a computer (digital).

For motion pictures, the technology exists to create, add, delete or change one object into another (morphing) with appropriate accompanying audio. Virtual additions are routinely used in professional sports broadcasts, such as the simulated banners on the walls behind baseball batters, ads on car racetracks and when the first-down lines on football fields are added to broadcasts. Players and referees even occlude the virtual line whenever their bodies cross it. Using the same technology, products are now often placed in motion pictures and television shows after they are shot. In 1999, the ability to digitally alter video in real time was even achieved. Video can be altered as it is shot and is now as malleable as photographs. The malleability of images allows the creation of realistic virtual worlds, which has helped fuel the $20 billion video game industry (Weingarten 2004: E4). Such virtual worlds are so realistic that a significant proportion of the $1.4 billion the Army spends on training simulations each year is spent on video technology (Soller 2004: E4).

Computers can now also alter video so that people appear to say things they never said. The process is used to dub motion pictures, but it also makes it far easier to create deceptive altered video. Previous methods had attempted to create a virtual model of a person's mouth and then used a computer to render digital images of it as it moved. A new method developed at the Massachusetts Institute of Technology uses artificial intelligence to teach a computer what a person looks like when they talk. All that is needed is two to four minutes of video of a person talking. The computer captures the full range of motion of the mouth and surrounding area. The computer can then express any word as a combination of those faces, up to 46 in one example, similar to the way different colors make up other colors. The computer then goes through the video, learning how a person makes every sound. The process appears lifelike for a minute or two, but then the person still appears to lack emotion. New versions, undoubtedly, will only improve until they are indistinguishable from reality (Cook 2002: A1). Combined with the capability to electronically paste someone's face on someone else's body, as was done in the 2000 motion picture *Gladiator* (Schickel 2005: R2), the possibilities for deception and propaganda are as vast as the creativity of the most devious of human minds.

The future will enable propagandists and deceivers to create perfect altered audio for perfect altered video for events that never occurred. In fact, faked video with the attendant faked audio may be just the next step on a path leading toward lifelike holograms. Just as Princess Leia sent a holographic message to Obi Wan Kenobi in the motion picture *Star Wars*, the US Army Research Laboratory predicts that they will be able to create holographic military forces with appropriate audio "someday soon" (Latimer 2001: 306). As soon as holographic images are possible to transmit, there is no doubt that someone will alter them for deceptive purposes.

Just as the ability to manipulate images has dramatically improved and spread, the ability to deliver altered images globally also has increased exponentially in the past few decades. Between 1989 and 1999, for example, the number of television viewers on Earth tripled to more than 1.2 billion. CNN International, for example, reaches more than 209 countries (Bowdish 1999: 32–3). The wire services, such as the Associated Press (AP) and Reuters, also ensure that stories are spread rapidly, while the Internet and cheap international telephone service also contribute to the rapid spread of information.

The Internet is an ideal medium for distributing propaganda or a deceptive story because it has open portals with no gatekeepers to evaluate the validity of a story. Operations that spread propaganda via the Internet have low entry costs, blur national boundaries, make it extremely easy to disseminate vast quantities of information globally, and are difficult to counter effectively because the source is often impossible to determine. Anyone can have a web page to disseminate propaganda and, with computers, anyone can make information, with images, graphics and text, look authoritative. Therefore, a bogus story can be inserted easily into one of the millions of Internet web sites. The difficulty is in attracting attention, but the more sensational the video, the more attention a story would attract.

Because of the ease of access and its global reach, the Internet is already the favored propaganda tool for many militarily weak groups. Al Qaeda; subcommandante Marcos, leader of a revolt in the Chiapas region of Mexico; Sinn Fein, the IRA's political wing; the Tupac Amaru in Peru; the Euskadi ta Askatusuna (ETA) independence group in Spain's Basque region; and the Somaliland Separatist Movement all use the Internet for propaganda purposes. Saddam Hussein also used to have a website, http://196.27.0.22/iraq, although anyone wanting a reply had to wait while the e-mails were printed in Jordan and sent 500 miles by road to Baghdad, since no one in Iraq had Internet access (Adams 1998: 166–9; Molander 1996: xv, 23). The American, right-wing militia movement has used the Internet to disseminate its message, especially after the 19 April 1995 bombing of the federal building in Oklahoma City (Molander 1996: 23). In the mid-1990s, someone hacked into the Department of Justice web site and replaced a page with a background of swastikas, a picture of Hitler, and the title, "Department of Injustice." The official CIA and Air Force sites were also hacked and replaced by anti-government propaganda (Adams 1998: 164).

Propaganda in the form of images that are too graphic for the mainstream media can be spread over the Internet. Images of hostages in Iraq being executed, for example, were distributed for the most part via the Internet. In May 2004, "Nick Berg," a hostage who was beheaded in Iraq, was the second most popular search on Google. Although hosting services usually shut down sites that display such horrible images as beheadings, the images are quickly downloaded and passed on long after the original site is closed (Smith 2004: A1, A12). A 2004 study found that 24 percent of adult Internet users or about 30 million people had seen graphic images of the war in Iraq online (Associated Press 2004d: Online). Such is the power of the Internet to distribute propaganda.

The traditional news media then often report that a story, especially if it is dramatic, appeared on the Internet, which gives the story even wider distribution. For example, during the 2004 Democratic Party's primaries, Matt Drudge reported on his web site that "several media outlets" were investigating allegations that front-runner John Kerry had had a two-year affair with an intern. Many television stations, including KABC in Los Angeles, then covered what Drudge had reported (Rutten 2004a: E1, E14).

The Internet has also affected the life span and distribution of newspaper stories. Once a story is in a newspaper's database, it is available for years online. Whenever an Internet search is made for the key words related to the story, the article will often be recycled, even if it has been discredited. Few users will be diligent enough to read any later corrections or retractions to learn that the story has been invalidated. Such a structure is perfect for disseminating propaganda and deceptive stories based on altered images.

The Internet also ensures that, no matter how unbelievable, invalid and lacking in empirical support a story is, people can be found who will believe it and spread it. As Neil Macdonald (Personal conversation 7 July 2005) has argued, when someone suggested an outlandish story in a village pub in the Middle Ages or around the campfire on the American frontier in the 1800s, the odds that someone else would believe them was low. If there was, for example, a one-in-a-thousand chance that anyone would be gullible enough to believe a story, and the pub or campfire only had 20 people present, the odds of someone else believing and spreading the story was extremely low. Today, however, with the Internet, even the most outlandish story can find supporters. Even if only one in 10,000 people would be foolish enough to accept a story, the Internet allows the teller to post their story, have it seen by millions over time and find those one-in-10,000 people who believe it. The Internet has greatly expanded the odds that those who devise deception operations and propaganda will find people to believe and spread their stories, especially those that appear credible.

Altered images are already common on the Internet, in motion pictures, advertisements, glamor and entertainment magazines, and are increasingly beginning to seep into hard-news media. Whatever the type of media, Charles Cooper, executive director of the US National Press Photographers Association (NPPA), warned, "What you see is no longer what you get. Photographs used to be the most accurate representation of reality, but you can't take that for granted anymore" (Brugioni 1999: 5). The story on the Internet of a young woman's lone motorcycle ride through the nuclear wasteland of Chernobyl attracted attention in 2004, even if it and the accompanying photographs were faked (Mycio 2004: A3). Nude images of stars, often a star's head on someone else's body, are on thousands of sites. Politicians also have not escaped the image-alteration explosion. Clinton peers through binoculars, oblivious to the fact that the lens caps are still on, while George W. Bush reads to a class, unaware that the book is upside down. Although most altered images on the Internet are easily identified as fake, many manipulated images are not labeled as having been altered and are extremely difficult to detect (Jaffe 2002: Online).

Manipulated images are also common in motion pictures and on television. Woody Allen's 1983 film *Zelig* had Leonard Zelig (Allen) appear in newsreels interacting with dignitaries from the 1920s and 1930s including Presidents Herbert Hoover and Calvin Coolidge, fighter Jack Dempsey, playwright Eugene O'Neill and Hitler. In 1994's *Forrest Gump*, through the use of altered images, the title character (Tom Hanks) appears as a Vietnam War hero in combat, a ping-pong champion, meets the Black Panthers and President John F. Kennedy, and shows his war scar to President Lyndon Johnson. Digital computer technology was also used in the motion picture to make Gary Sinese, who plays a lieutenant who lost his legs in Vietnam, appear legless (Brugioni 1999: 57–9). More recently, the 1997 motion picture *Contact* altered two President Clinton news conferences, one about a Mars probe and the other about the Oklahoma City bombing, to make it appear as if he was talking about alien communications. Their use drew criticism from the White House (Brugioni 1999: 197). Computers now allow the creation of entire worlds that do not exist, such as in 2004's *Sky Captain and the World of Tomorrow*. The film flopped, but did show how CGI could create entire worlds. Even the dead can be resurrected for one final scene. In 2000, Nancy Marchand, the actress playing Tony Soprano's mother, Livia, on the popular HBO series, *The Sopranos*, died when the show was on hiatus. The producers wanted a final scene with Tony and his mother, so they used old images of Marchand as Livia to create a new, final scene. In a similar fashion, Sir Laurence Olivier, long since dead, was used as the villain in *Sky Captain and the World of Tomorrow*. Reality television, such as *Survivor, The Apprentice* and *Big Brother*, has reached a new level of manipulating images by staging and altering the order of scenes, and adding sound effects or voice-overs to distort the meaning of scenes.

While most motion picture manipulations, of course, intend no harm, those done in a quasi-documentary format, such as Oliver Stone's 1991 controversial conspiracy film, *JFK*, blur the line between entertainment on the one hand, and propaganda and deception on the other hand. The History Channel even used a documentary format to argue that then Vice President Johnson was involved in Kennedy's assassination. The documentary format made the claim appear valid even though it is unsupported by any credible evidence or espoused by any reputable historian. The History Channel later apologized for the allegations and said it would not re-air the show or offer it on videotape (Margulies 2004: E2; Shaw 2004: E14). More recently, in 2004, Michael Moore's *Fahrenheit 9/11* was presented as an alternative view of the years preceding the 2003 US intervention in Iraq. If judged as a documentary, the film had logical and factual weaknesses in large part because it focused on certain things and not others. As propaganda, however, it was extremely powerful. When criticized for its inaccuracies, Moore could claim that the film was just entertainment, even as he admitted it propagandized his own views.

Advertisements have also seen an explosion in the use of altered images. In the 1980s in Britain, footage of Gary Cooper from *High Noon* was used in a Holstein lager television advertisement, while John Wayne from *Cast a Giant Shadow* was used by Coors to promote their beer. Diet Coke used images of Humphrey Bogart, Louis Armstrong and James Cagney in the new drink's first television advertisements. In spots for the 1992 Olympics, Gene Kelly danced with a living Paula Abdul, Cary Grant poured her a Diet Coke and she conversed with Groucho Marx. In 2004, Steve McQueen, who died in 1980, starred in a series of television ads for the Ford Mustang. DHS vans fly, Budweiser Clydesdales play football in the snow while zebras officiate, the Merrill Lynch bull is scratched behind the ears by a couple in their living room, and all manner of the unbelievable and undoable appear to be completely realistic. As with motion pictures, however, with most advertisements there is no intent to deceive or to cause harm.

Altered images are common in tabloids, some newspaper sections and in many types of magazines, especially those that cover entertainment and fashion. Models are altered to appear to be perfect. It is standard to remove wrinkles, lighten hair, tuck tummies, trim thighs, increase cleavage and shorten dresses in what has become digital plastic surgery. Such techniques are called Scitexing, after Scitex, the leader in the image-manipulation field (Brugioni 1999: 15). Sales drive the need to alter, especially on magazine covers, which are used to attract buyers. Such manipulation is done even more in Britain, for example, where about 80 percent of magazines are bought off newsstands than in Canada where 80 percent of magazines are bought by subscription (Cobb 2003: Online). Often tabloids create images if no photograph is available. For example, a photograph of Nicole Brown Simpson in the *National Enquirer* appeared to show her after being allegedly beaten by her ex-husband, O.J. Simpson. Only the small print explained that the image was a computer recreation of how her sister described the injuries (Brugioni 1999: 14).

Some educational magazines have also used altered images. The most famous case occurred in February 1982, when *National Geographic* moved the pyramids of Cheops and Mycerinus so they would fit better on their cover. The magazine called it "retroactive repositioning of the photographer" (Brugioni 1999: 134). Many historical television programs often mix film of an actual event with film from other, similar events as illustrations, without identifying them as such.

A story based on altered images does not necessarily have to find its way into the hard-news arena to have a powerful effect. As journalist Marvin Kalb said, "More Americans get their print 'news' these days from *People* Magazine, the *National Enquirer*, *TV Guide*, or the *Star* tabloid than they do from *The New York Times* or *The Washington Post*" (Brugioni 1999: 53). Even while many pundits dismiss the tabloids and many blogs as lacking credibility, large segments of the public do not distinguish between the hard-news pages of a newspaper with their strict rules against altering images and the color supplements, glossy inserts and soft-news feature stories, where rules for manipulating images are far less stringent.

Although previously in the United States a partisan press often stretched the truth or outright lied, both in text and with images, the modern hard-news media has sought to achieve credibility through objective reporting. In news reporting, altering images without informing the audience is considered deception and an ethical breach on par with inventing stories and presenting them as fact. The *Chicago Tribune*'s director of photography, Jack Corn, called electronic manipulation, "Ethically, morally and journalistically horrible." Stephen Isaacs, acting dean of the Columbia School of Journalism, told the AP in 1994 that "To distort reality is a journalistic sin." The NPPA mandates that "Altering the editorial content of a photograph, in any degree is a breach of the ethical standards recognized by the NPPA" (Brugioni 1999: 12). If an image is altered, most news organizations argue that it always should be labeled as such.

Despite their guidelines against manipulating images, many news-media outlets have used altered images. In news photographs of accidents, blood is often softened or removed, and facial expressions on victims are sometimes changed (Brugioni 1999: 11). After his arrest on suspicion of murdering his ex-wife and her friend in 1994, *Time* magazine darkened O.J. Simpson's cover photograph. The photograph was labeled a photo illustration, but *Time* was still widely criticized (Brugioni 1999: 13; Frank 2003: B11).

Even news coverage of war, national security and politics has involved altered images. The October 1987 cover of *Newsday* showed an image of 18 F-14 jets taking off in formation. The image was created by taking a photograph of one jet and copying it 17 times. During the 1990–1 Gulf War, the gray sky during the Battle of Khafji was altered into a brilliant blue on *Time*'s cover (Brack 1996: 48; Brugioni 1999: 9). In May 1996, the *Evening Standard* of London apologized to John Prescott, MP, over its manipulation of a photograph of him and his wife at a party. The beer in front of Prescott had been airbrushed out, while the bottle in front of his wife had been cropped to make it appear to be a champagne bottle. The caption: "Champagne socialist: Prescott in dinner jacket with wife Pauline and, left, [in another photograph] his Jaguar." It was the first instance known to Britain's National Union of Journalists of a photo manipulation for a political purpose (Freelance 1996: Online).

More recently, on 31 March 2003, the *Los Angeles Times* published a photograph of a British soldier directing Iraqi civilians to take cover during a battle near Basra, Iraq. Brian Walski, the photographer, however, had used a computer to combine images from two photographs taken moments apart "to improve the composition." The newspaper fired Walski for violating their policy that forbids altering the content of news photographs, and then printed an explanation, as well as the two images used to create the photograph (Los Angeles Times 2003: A1, A6). Such alterations, however, are far from unique and often not so publicly disavowed. On 9 April 2003, the front page of London's *Evening Standard* showed a large and enthusiastic crowd cheering Saddam Hussein's overthrow. Repeating patterns in the image suggest that the crowd was much smaller, but was copied and repeated. Some people in the crowd appear to be in it three times. The newspaper explained that it was an image from BBC television and that they only removed the BBC logo and replicated part of the crowd to fill in the space left vacant by the deleted logo (Camera/Iraq.com 2003: Online).

The use of altered images is even worse in many other countries. In Taiwan, for example, the highly competitive news business has led to reporters staging scenes to illustrate stories if they lack real footage (Magnier 2005: A6). In India, a major male Bollywood star claimed that a video of him soliciting sex from a young journalist posing as an aspiring actress in exchange for a role, had been altered (Badam 2005: E11). Altered video has probably appeared on some US newscasts, especially since the images are on screen only briefly, making it extremely difficult to spot any manipulations. Eric Haseltine, Senior Vice President for Research and Development at Walt Disney Imagineering, said, "I'm amazed we have not seen phony video. Maybe we have. Who would know?" (Amato 2000: Online).

Given the potential of altered images, which has been so extensively demonstrated by the media, the United States and other media-saturated countries are prime targets for propaganda and deception operations that use manipulated images. Therefore, the study of the principles and tactics of propaganda and deception is crucial to recognizing such attacks and developing the means to counter them.

3

PSYCHOLOGICAL OPERATIONS

Hearts and minds and eyes

Throughout history the support of the people has been crucial to military operations and to government policies, although that support became even more important with the rise of nationalism in the late 1700s. By the twentieth century, hearts and minds were critical to victory, with General Dwight D. Eisenhower concluding, "Public opinion wins war" (Knightley 1975: 315). Therefore, psychological operations (psyops) seek to convey selected information to target audiences to influence their behavior and, through them, government policy. Given the power of images, altered images can play a crucial role in psyops.

The most common type of psyops is propaganda, which is the use of any form of communication to spread or reinforce certain beliefs for political purposes. The word "propaganda" originated in 1622 during the Counter Reformation as the name of the Congregation of the Roman Curia or council charged with spreading the gospel. Originally the word "propaganda" was neutral, but with the rise of the Nazis and Communists, it has taken on a negative connotation. Like advertising, propaganda is a tool, however, that can be either positive or negative. It ranges from Leni Riefenstahl's 1934 *Triumph of the Will* that glorified Hitler and the Nazi Party to Frank Capra's *Why We Fight* series that supported the US war effort during the Second World War.

The most effective propaganda combines entertainment, education and persuasion. The entertainment elements attract the audience, while the educational aspect decreases the perception that the message is propaganda, even as it persuades. Unlike education, which seeks to present an objective view, propaganda is biased through the selective use of facts, although the bias is usually subtle. Sometimes the difference between propaganda and education is just *post hoc* rationalization: if a persuasion attempt fails, it is called propaganda; if it succeeds, it is called education.

Ideally, propaganda is developed after a policy goal has been clearly defined. It is best if psyops specialists are involved from the beginning of any operation, so psychological factors can be taken into account in the early planning stages. Then analysts conduct research on, and gather intelligence about, the target audience before developing the propaganda itself. If possible, the propaganda is then pre-tested, such as on cooperative prisoners of war (POWs) during wartime or on focus groups of individuals who are similar to the target population during peacetime. After revision, the propaganda, in whatever form, is produced and disseminated to the target audience. Feedback from any number of sources, such as newly captured prisoners during wartime, spies, moles or analysis of the target state's media, is then used to evaluate the effect of the propaganda. This analysis is used to refine the propaganda to try to make it more effective.

The development and production of propaganda is usually decentralized in order to increase its local flavor and to avoid making any cultural or linguistic gaffes. Despite this, there is a great need for a state to centralize coordination of all propaganda to ensure a consistent message that supports economic, diplomatic and military policy. If coordination between propaganda and other government policies is not close, then the propaganda effort is often wasted. For example, in the early 1960s, the United States gathered information on Sukarno's "copious sexual misdeeds" and spread the ribald tales through channels not attributable to the United States. At the same time, however, Washington blocked Holland's request to allow Dutch warplanes on their way to prevent Sukarno's conquest of West Borneo to refuel at US bases. Propaganda and diplomatic policy were at odds, and Indonesia took West Borneo, thus increasing Sukarno's political stature (Codevilla 1989: 82–3).

Depending on the source, propaganda can be white, black or gray. In white propaganda the source is known and is usually official; in black propaganda it is concealed and a false source is suggested; and in gray propaganda the source is obscured. Black propaganda is more difficult to create than white propaganda because white propaganda can contain mistakes and still be effective since it is known to be from a foreign source, whereas black propaganda cannot contain certain types of errors. For example, during the Second World War the British operated a radio station that masqueraded as being operated by loyal Germans broadcasting from inside occupied Europe. While Germans might expect an official British communiqué to sometimes contain inaccurate information about Germany that any German would know, the same error from a station pretending to be German would cripple the station's credibility. Although black propaganda may engage the public's imagination far more than white, Richard H.S. Crossman, British Deputy Director of the Political Warfare (PW) Department for the Supreme Allied Expeditionary Force (SHAEF) during the Second World War, said:

> All I would say about the "black" [propaganda] is that, on the whole although we found [its] activities enormous fun, although a vast amount of talent went into them, although I am sure they entertained the Gestapo, I have grave doubts whether "black" propaganda had an effect in any way commensurate with that of ordinary straightforward [white] propaganda.
> (Daugherty 1958a: 37)

In gray propaganda, the link to the source is often discernable, but may not be clearly proclaimed. The British used gray propaganda extensively during the

Cold War. By the 1950s, the Information Research Department (IRD), which ran British gray propaganda, was the largest department in the Foreign Office with a staff of 500. IRD propaganda, which regularly secured lead stories on the BBC that were then syndicated on the front pages of newspapers around the world, was extremely effective, largely because IRD stories did not appear to be propaganda. They were well researched, factual and provided free of charge, so journalists in many countries were happy to use them. The IRD was also arranged into country desks, so the stories had a local flavor, and the British fed the stories into the media through local sources to obscure their British origin (Aldrich 2002: 445, 457).

The perceived intention of a source significantly influences the effectiveness of propaganda. If a source is perceived to be a propagandist attempting to manipulate an audience, then effectiveness plummets dramatically. Crossman said, "The art of the propagandist is never to be thought a propagandist, but seem to be a bluff, simple, honorable enemy who would never think of descending to the level of propaganda." In one sense, Crossman said, "One must hate propaganda to do it well" (Daugherty 1958a: 39). While the Nazi radio commentator Lord Haw-Haw was thought to be a spectacular success by his creators, his intended British audience perceived him as an unabashed propagandist and his attempts to influence them failed. Conversely, the British were far more effective at propaganda because the operators they selected to run propaganda operations disliked it and went to great lengths to conceal what they were doing. Many operators, in fact, disliked it so much that after the war almost none of them discussed their wartime propaganda roles with historians, let alone in memoirs. As with the Nazis, most Communist propaganda aimed at Western audiences during the Cold War was ineffective because it sounded like propaganda. In West Germany, for example, while only 9 percent of the population listened to the blatantly propagandist East German broadcasts, 65 percent listened to the Voice of America (White 1958: 618). Although both stations were designed to disseminate propaganda, the subtle tone and high news and entertainment value of the Voice of America attracted a far larger audience.

A source is more effective if the target audience appears to be an unintended recipient of information directed at a different audience. In such a situation, the source does not appear to be seeking to persuade the target audience and appears to be more credible. Reprinting articles from other US magazines, for example, made *American Illustrated*, a magazine for Soviet citizens during the Cold War, more credible, since Soviet readers were getting information that was also sent to US readers and not just propaganda designed for them (Dasbach 1976: 720–2).

A source that is similar to the target audience can increase the credibility and, therefore, the effectiveness of propaganda. For example, during wars, cooperative POWs are often used to broadcast to their former comrades to attempt to persuade them to surrender. The key is to ensure that the selected POW is still similar enough to his former comrades to be effective. In one case in the Marianas during the Second World War, US troops captured a Japanese soldier named Kobayashi.

Having concluded that the war was lost, Kobayashi agreed to use a loudspeaker to try to convince other Japanese soldiers to surrender. He was extremely good at it and became so trusted that the Americans let him go out on his own to make his broadcasts. Then, at the end of a long, hot day, one of the US officers invited Kobayashi to his tent for a beer. Soon Kobayashi was playing cards and socializing with the Americans. From then on his effectiveness as a propagandist dissipated. He no longer saw the world the same way as his former comrades and his attempts to convince other Japanese soldiers to surrender consistently failed (Cotler 1958: 198–224).

As with the law, for which Clarence Darrow said, "The main work of a trial attorney is to make the jury like his client" (Cialdini 1995: 268), sources are also more effective propagandists if the audience likes them. An audience tends to like a source if the source is good-looking, or if the source compliments, cooperates or helps the audience, such as by providing supposedly secret information (Cialdini 1995: 262–4, 269–70). Mere exposure, however, can also increase liking, which forms the basis for the propaganda emphasis on repetition of a message. In fact, even if people are shown something as simple as a pen several times, their liking for the pen will increase the more often they see it (Tesser 1995: 223). The influence of the source, however, decreases over time. In what is called a sleeper effect, as time passes, people forget the source, regardless of its credibility, and just remember the message. Even if information is received and not believed because the source lacks credibility, the person who received the information may later relate that information to a friend as if it is true, having forgotten the source (Myers 1996: 277).

Audiences are far more likely to be persuaded by sources they deem to be credible, especially if the source is perceived to be an expert in the topic area. As Virgil counseled, "Follow an expert." Credibility is often tied to prestige. Audiences are more likely to believe the Chairman of the Joint Chiefs of Staff (JCS) on military matters, for example, than a high school history teacher. Credibility, however, is only partly an attribute of the source itself. The audience also determines a source's credibility. For example, to Americans the Chairman of the JCS would be highly credible on military affairs, but he would be far less credible to the public in Iran, Cuba or China.

The most effective means of appearing credible is to tell the truth. Crossman said:

> In propaganda truth pays.... It is a complete delusion to think of the brilliant propagandist as being a professional liar. The brilliant propagandist is the man who tells the truth, or that selection of the truth which is requisite for his purpose, and tells it in such a way that the recipient does not think that he is receiving any propaganda.... If you give a man the correct information for seven years, he may believe the incorrect information on the first day of the eighth year when it is necessary, from your point of view, that he should do so. Your first job is to build the

> credibility and the authenticity of your propaganda, and persuade the enemy to trust you although you are his enemy. . . . The art of propaganda is not telling lies, but rather selecting the truth you require and giving it mixed up with some truths the audience wants to hear.
>
> (Daugherty 1958a: 39)

For example, although the demand for unconditional surrender posed a significant problem for Western propagandists during the Second World War, it also helped Western credibility immensely, since it meant that the West made no promises that could later be broken (Daugherty 1958a: 42).

Conversely, one of the most difficult times for British and American propagandists during the Second World War was when, in 1943, the Nazis told the truth after they discovered the corpses of more than 10,000 Polish soldiers, mostly officers, in Katyn forest, near Smolensk. This information put the Western allies in an extremely difficult position. If they denied the allegation and information surfaced that the Soviets had committed the atrocity, it would have severely damaged Anglo-American credibility. On the other hand, the Western allies certainly did not want to support Nazi allegations against the Soviets. The only response left, weak as it was, was to focus on the intent of the source and point out the propaganda objectives of the Nazis (Janowitz and Marvick 1958b: 355–6). This episode not only highlights the difficulty of responding to truthful propaganda, but also the difficulties that sometimes arise in alliances, especially between democratic and totalitarian states.

During the Vietnam War, the North Vietnamese also realized the importance of credibility and its relationship to the truth. They accurately reported the devastation caused by air strikes, albeit focused on damage to civilian targets, which both strengthened domestic resolve and evoked international sympathy (Quang 1976: 1115–19). Credibility and truthfulness were also crucial to US propaganda efforts during the war. When leaflets argued that the VC should defect to the government or die under a torrent of B-52 bombs, many VC believed it because they knew the power of B-52 bombing raids all too well. On the other hand, Vietnamese government leaflets that claimed the South Vietnamese people were enjoying a prosperous, free life were ineffective because the VC knew that such claims were false. Many of the southern people were miserable, had little or no freedom, and were undergoing intense bombings (Pappas 1976: 277–9).

Telling the truth even when the news is bad for your side greatly enhances credibility. "From the point of view of psychological warfare," Crossman said, "a defeat is a great opportunity, especially if you . . . say that your defeat is worse than it is. You must be frank about it, franker than the facts. Then you really begin to gain the enemy's confidence" (Daugherty 1958a: 40). For example, *Frontpost*, a US-produced newspaper for Germans during the Second World War, accurately reported Allied military operations even during the Battle of the Bulge when the Allies were retreating. By reporting an Allied setback, the newspaper's credibility soared (Daugherty and Janowitz 1958: 559). Accurate reporting of the news, even information that appeared to help Germany, also caused the BBC to be regarded as being both truthful and credible. After a British air raid, for example, the BBC reported that seven Royal Air Force (RAF) planes had failed to return, while the Germans announced that only five RAF aircraft had been shot down. The Germans had been unaware that two damaged aircraft crashed into the North Sea. Crossman said, "That was the greatest psychological warfare triumph of the year!" (Daugherty 1958a: 40). British reporting also avoided sounding like propaganda by downplaying their successes. Crossman argued, for example, that a wonderfully successful air raid should be called only a moderate success. "That sort of thing gives the enemy cold shivers" (Daugherty 1958a: 41).

The air-raid example highlights the paramount importance of being accurate if an audience has access to other sources of information, especially today when audiences often have access to newspapers, magazines, radio, television, Internet websites, e-mails, faxes and cheap international telephone service, all of which facilitate the rapid spread – and checking – of information. If part of a message can be checked against other sources and is found to be wrong, credibility will suffer. Conversely, if part of a message can be checked and found to be accurate, the audience is more likely to believe the entire message. Furthermore, if the audience receives the same message from multiple sources, they are then more likely to believe the message.

Unlike libel, where truth is an absolute defense, truth in propaganda is far from a guarantee that the intended audience will find it credible. Credibility resides in the minds of the audience. Therefore, a propaganda message must be credible to the intended audience, regardless of whether it is factual or credible to the originator of the message. For example, during the Korean War, US General Matthew B. Ridgeway wanted more use of images in propaganda leaflets, especially images that highlighted the horrible devastation that United Nations (UN) firepower could cause on a battlefield. It was difficult, however, to find usable photographs, since most post-battle images had GIs, often grinning with relief, prominent in the foreground and the enemy dead only in the background. Psyops officers finally found a usable photograph with the caption, "Battlefield scene showing Chinese Communist dead on Korean hill 262, after successful assault by Turkish RCT [Regimental Combat Team], February 21, 1951." The psyops officers believed this was an excellent photograph, since besides showing the power of UN weaponry, it would also show the Chinese that the war was still being waged by a UN coalition, not just by the United States and their South Korean allies.

The psyops officers printed leaflets with the photograph on the front and a message on back: "These Chinese needlessly met death resisting UN Turkish forces," with information about how to surrender. Then the officers used a panel of cooperative Chinese POWs to evaluate the leaflet. All of the prisoners agreed that the image was excellent, as was the text. The psyops officers were feeling good about their work and the meeting was about to end when one POW asked, "But, why did you spoil a good leaflet by telling a lie?" The POW explained to the bewildered psyops officers, whose pamphlet was based entirely upon facts, that there could not be any Turks in Korea, since all of the Turkish soldiers had been killed north of Pyongyang in November 1950. All of the Chinese POWs nodded agreement that this was true. In fact, on the night of 25–26 November 1950, the Chinese did maul the Turkish RCT and the battle received a great deal of press attention. The unit, however, had been reformed and was back in the line fighting, but the reformation had received little attention in the Chinese media. Apparently the Chinese soldiers had not heard about it and assumed that there were no Turks left in Korea (Daugherty 1958e: 612–14). As Crossman warned, "You have to eliminate a great many things which are true if the enemy will not believe them to be true" (Daugherty 1958a: 41).

Conversely, the credible, even if it is not true, can be made to appear truthful and can harm an adversary. In the summer of 1941, for example, when the Germans were advancing on Kiev, the British broadcast that if the Germans could not reach the Ukrainian capital in 10 days then their whole campaign would falter. The British had determined that it would take the Germans 12 days to reach Kiev and were therefore able to make the German advance appear to be a failure (Daugherty 1958a: 40).

The personnel required to conduct a psyops campaign are many and varied. During the First World War the British recruited novelist and futurist H.G. Wells to run the German unit, and Wickham Steed of the *Times* and academic Hugh Seton-Watson to focus on Central Europe and the Balkans (Smith 1989: 40). During the Second World War, Theodore Geisel, the future Dr Suess, and the great motion picture directors John Ford and Frank Capra, among many other artists, were involved in US propaganda work. A US psyops team in Assam in South East Asia during the Second World War "included newspaper reporters, advertising men, missionaries, artists, radio engineers, teachers, and mechanics....a Shan prince, a British bishop, a Rangoon lawyer, Indian Moslems and Hindus of various castes, Chinese radio monitors, Japanese-educated Chinese, and a Kachin agriculturist" (OWI Psychological Warfare Team 1958: 185).

Even with such varied personnel, however, most psyops operators have backgrounds in newspapers, magazines, radio, television, public relations or advertising. British propaganda during the First World War, for example, was run by two newspaper barons: the Canadian-born William Maxwell Aitken, who would become Baron Beaverbrook, and Alfred Charles William Harmsworth, Lord and then Vicount Northcliffe. In the Second World War, the British Morale Operations (MO) branch, which created black propaganda, was staffed by journalists, authors, screenwriters, and advertising and public relations experts.

Although they share many traits, it should be remembered that many journalistic and advertising techniques are not applicable to propaganda. Exhortation tends to stiffen an enemy's morale, while a sales talk will be seen as such and will fail. The key to propaganda is to repeat the message over and over, and over yet again. Journalists usually want to provide news that is new, which conflicts with propaganda's requirement to repeat a story or theme *ad nauseum*. Repeating the same news day after day, for example, was agony for journalists who worked for the British propaganda services during the Second World War (Daugherty 1958a: 39).

Several different types of personnel are required to operate a psyops organization. Idea men, who provide the basic idea for the themes and are usually few in number, should be extroverts with keen imaginations, unrelenting perseverance and an eagerness to convey ideas. They should also be adaptable and patient. They should have linguistic and cross-cultural skills, some military experience and area knowledge, often in the form of educational, journalistic or business experience in the target nation. During the world wars, for example, some of the most effective Allied propagandists in the European theater had been raised or educated in Germany. Part of their effectiveness was empathy for the target audience; not sympathy, but empathy in the sense of being able to think and see the world in the same way as the target audience. Crossman, for example, a prewar Oxford professor, had "an imagination to see the machinations of the enemy" and possessed an "intimate knowledge of German and of the German character" (Daugherty 1958a: 35).

Psyops also requires administrators, who should be well-traveled, experienced executives, and often have a background in publishing, radio, television, advertising, public relations and, today, the Internet. They are usually not specialists. Psyops teams also include researchers with backgrounds in psychology, sociology and political science, and technicians who actually produce the propaganda material, operate the radio or television stations or create websites. Liaison men are also needed to run operations on the ground. They should show initiative, be able to get along with all kinds of people, and be familiar with psyops and the local language. Psyops teams also have creative personnel, such as writers, artists, photographers, cameramen, and announcers who actually create the messages. They should know the target, usually by having lived abroad, be fluent in the language, and be able to express themselves clearly and accurately. They also need to be flexible enough to write or draw in the local style.

Sergeant Benno Frank was one of the top US propagandists in Europe during the Second World War and was an outstanding example of a psyops operator. Educated in Germany, he had an extensive background in German culture and could easily empathize with German views of the world when he interrogated German POWs and made propaganda broadcasts. He was also a character in the finest traditions of the extroverted, creative and maverick psyops operator. Trained in the theater, Frank's broadcasts were effective and entertaining. He "knew when to shout and when to whisper" (Daugherty 1958b: 249). Frank never wore his sergeant's stripes and referred to his division commander by his first name. He also quickly convinced his superiors to allow interrogators to take POWs out of prisoner cages to interview them under more relaxed conditions.

In his broadcasts, Frank was known to the Germans as Captain Angers, a soldier who had served in the German Army and later became a US Army captain. The story, while incredible to Americans, was credible to Germans. His repetitive theme was that it was easier and more cowardly to continue fighting the Allies and die on the battlefield than to surrender and live to face heroically the immense task of rebuilding Germany. Frank once broadcast, "Come over. If you don't like it here after a 30-hour trial period, you will be free to go back. On my honor, I will see to it that you are sent back! Ask for Captain Angers." One day, Fridolin Hopt, a captured ardent Nazi, demanded Angers keep his promise and return Hopt to the German lines. Frank faced a dilemma. His offer had not been cleared with his superiors and the US Army generally takes a dim view of returning captured enemy soldiers to their own lines to fight another day. Frank, however, not wanting to harm Captain Angers' credibility, convinced his superiors to release Hopt. Frank then used Hopt's name and unit in later broadcasts to prove that Captain Angers was a man of his word. After that, the Germans probably would have believed anything he broadcast (Daugherty 1958b: 250).

After studying First World War propaganda, a British Member of Parliament (MP), Arthur Ponsonby, wrote;

> You must have intellectual lies for intellectual people and crude lies for popular consumption, but if your popular lies are too blatant and your more intellectual section are shocked and see through them, they may…begin to be suspicious as to whether they were not being hoodwinked too. Nevertheless, the inmates of colleges are just as credulous as the inmates of the slums.
>
> (Ponsonby 1928: 24–5)

Ponsonby was onto something: there do appear to be two routes to persuasion. The central route involves the evaluation of arguments, as individuals think them through, while the peripheral route involves people changing their attitudes without much thought, such as deciding which brand of soup to buy (Tesser 1995: 207–8). More educated, intelligent people are more influenced by arguments based on reason via the central route, while disinterested audiences are usually more swayed by emotional appeals via the peripheral route (Myers 1996: 281). For this reason, propaganda often includes both rational arguments and emotional appeals.

Whether devising a rational or emotional message, propagandists face a choice: make a one-sided presentation solely with information to support their desired views or a two-sided argument that presents information about both sides of an issue. If an audience is strongly opposed to a message, as in many foreign policy and war situations, propaganda is more effective if it offers both sides of an argument. Even if the audience is not opposed, however, those who hear two-sided arguments are, in a form of inoculation, more likely to resist later counter-persuasion. In fact, even if an audience is only warned that an attack on a belief they hold is coming, they are less likely to be persuaded by later propaganda attempts (Janis 1976: 614).

In a two-sided presentation, however, which argument should be presented first? In what is called a primacy effect, if an audience is not familiar with the opposing arguments, then it is more effective if arguments in favor of a position are given before the opposing arguments. If an audience is familiar with the opposing arguments and doubts the communicator's honesty, however, then it is more effective to present the counter-arguments first and then arguments in favor of the communicator's position. This later scenario is more common in times of war, and in political and foreign policy situations (Janis 1976: 615). Shakespeare has Marc Antony do this when he makes his speech to the mob after Julius Caesar is assassinated. Antony begins not by challenging the mob's belief that Caesar was a leader deserving of death, "I come to bury Caesar, not to praise him," but by reinforcing their belief that the assassins were in the right, "For Brutus is an honorable man; So are they all, all honorable men" (II, III, 76, 84–5) – at least at first, before making his arguments damning the assassins.

Propaganda can also make an implicit or explicit appeal. If a topic is well known, it is more effective to allow an audience to draw conclusions for themselves, and not state the conclusion explicitly. If a topic is less familiar to an audience, then the propagandist should explicitly state the conclusions (Janis 1976: 617). In 1940, for example, almost all Britons realized that they faced Nazi invasion and possible defeat, so the BBC broadcast implicit propaganda by announcing plans for a series of talks on Charlotte Bronte. The planned talks boosted British morale, since the implicit message was that Britain would not be defeated and normal life would continue. Similarly, information broadcast about criticism of Churchill was powerful propaganda in Germany, since it highlighted, implicitly, what the Germans could not do: criticize Hitler (Daugherty 1958a: 46).

Such implicit propaganda can even be largely unintended. As Neil Macdonald (personal conversation 21 August 2005) argued, television programs, sports coverage, motion pictures, books and many other art forms often contain implicit propaganda. In television programs, motion pictures and sports coverage of the Olympics, for example, each nation almost always shows its own people as heroes and winning, which provides the implicit message that they are superior. The motivation is far from sinister, usually just to put on a show that the public in each nation will watch, but the implicit message of a superior people can be powerful propaganda.

Propaganda containing messages that an audience may dislike is often more effective if the messages are presented implicitly within a neutral or even positive framework. Both sides used this technique during the Cold War. Between 1960 and 1963, for example, *USSR* magazine, which was produced for a Western audience, decreased the number of stories on government and foreign relations, and increased the number of articles on people and science to present a more humane, learned image of the Soviet Union. Similarly, the 1963 *America Illustrated*, targeted at a Soviet audience, increased stories about the arts and culture to make the United States appear more cultured and counteract the image that America was a cultural wasteland (Dasbach 1976: 720–2). Both magazines sought to project an implicit image of their respective countries through a neutral framework, because an explicit message might have been rejected outright as propaganda.

A common propaganda technique is to use attractive generalities to describe the propagandist's position, such as that they support life, liberty and the pursuit of happiness, with no explanation of what any of the generalities mean. In this way, simplification is used to reduce propaganda themes to short, simple slogans. Celebrating a battle against native Americans in 1811, "Tippecanoe and Tyler, Too," helped win the 1840 US presidential election for Tippecanoe victor William Henry Harrison, although it benefited him little, since he died a month after catching a cold at his inauguration. During the Second World War, British agents scrawled "For this we have the Fuhrer to thank" on bombed-out buildings in Germany (Becker 1958: 674). The VC also used slogans but, given the rural nature of Vietnam, they instead carved slogans on the bark of tree trunks. When government soldiers spotted the carved slogans, they wanted to cut down the trees. The villagers, however, wanting to keep the trees, asked, "If the Liberation soldiers had written the slogans on bridges would you blow up the bridges?" The soldiers were forced by this logic to withdraw without cutting down the trees, leaving the slogans intact (JUSPAO Planning Staff 1976: 682–3). In 1964, Barry Goldwater ran for the US presidency with a slogan "AuH2O" as a play on his name. His opponents were soon scrawling "Gold for the rich, Water for the poor" on fences. More recently, President Ronald Reagan used "Trust, but verify" to encapsulate his philosophy for dealing with the Soviets.

Propagandists also use repetition. Even with no supporting information, repetition increases the believability and acceptance of a statement. It has been often repeated, for example, that Reagan was the most popular president in US history, yet his approval rating averaged 50 percent, lower than the averages of Eisenhower (69 percent), John F. Kennedy (71 percent), Johnson (52 percent) and Nixon (56 percent), and not much above Jimmy Carter (47 percent). Reagan's highest rating, early in his first term, was 68 percent, lower than that of the five previous presidents (Fairness and Accuracy in Reporting 2004: Online). Repetition, however, has turned the myth of Reagan's popularity into a fact in the popular mind.

A propagandist can also appeal to a religious, political, or ideological authority to legitimize their views. For example, states often seek United Nations' support for military interventions in an attempt to transfer the UN's authority and legitimacy to their own actions. The opposite method is to use a regular person as a spokesperson, with the assumption that the target audience will think the source is like them and, therefore, more credible. The image of President Carter as just a small-town, peanut farmer from Georgia relied on this concept, even though he was in fact a highly educated US Navy nuclear engineer. Benno Frank's Captain Angers used the same concept of an average, albeit German-born, American officer broadcasting to German troops.

Another propaganda technique is to take advantage of the tendency to stereotype people. A stereotype is a summary impression of a group of people in which a person believes that all members of that group share a common trait or traits. A stereotype can be positive or negative, such as viewing your own soldiers as courageous, chivalrous and good, while believing the enemy is cowardly, brutal and evil. Derogatory names or images also can be used to dehumanize a group of people, thus making it easier to persecute them. Hitler, for example, called Jews and Slavs "*untermenchen*" or sub-humans. Scapegoating is often combined with this method, in which a group or another state is portrayed as the cause of a nation's problems. Scapegoating helps strengthen feelings of unity and solidarity within groups, and cements an us-versus-them mentality. Bandwagoning, when a propagandist attempts to give the impression that everyone thinks the same way about a topic, can also be tied to scapegoating. The goal is an illusion of universality; such as during the Cold War, when some Americans promoted the idea that "everyone" knew the Soviets could not be trusted.

Whatever techniques are used, the most effective propaganda is paired with other methods, such as the spreading of rumors, information sent through double-agents, deception operations, diplomacy and/or military operations. Altered images can also play a significant and powerful role in psyops and propaganda, either as the primary tool or as a supporting technique.

Apart from deciding which techniques to use, propagandists must also select themes to emphasize. Just as with credibility, the effectiveness and validity of a propaganda theme rests in the minds of the audience. In the Second World War, for example, the Japanese did not have an accurate understanding of propaganda themes that would be effective against the United States. In 1943, for example, a Japanese radio announcer told Americans:

> In your country you have a national case of pink-toothbrush on a large scale and in many forms. The American way of living made most comfortable and easy by advanced science is not all it is cracked up to be. Survival of the fittest, remember.... After twenty years of soft living the American soldiers are no match for the Japanese.
>
> (Berreman 1958: 434–5)

Another broadcast explained that the Japanese soldier was superior because of "mental energy" (Berreman 1958: 434–5). Although these arguments made sense to the Japanese, to Americans, who perceived materialism and high living standards as the basis of US culture and power, the attack on US materialism was doomed to failure and talk of "mental energy" was dismissed as Japanese mumbo-jumbo, totally without credibility. Similarly, during the Vietnam War, VC attempts to focus on political themes and promises of sex if American troops defected also failed, given that most US soldiers in combat did not see their immediate problems as caused by politicians in Washington, but rather by their VC opponents. Sex was also far from uppermost in their minds as they attempted to survive their tours. The VC also praised their own troops with headlines that insulted Americans, such as "US Troops' Shameful Withdrawal from Khe Sanh Base," which, hardly surprisingly, failed to persuade American soldiers to surrender (Herz 1982: 351–5).

Propaganda messages that invoke fear can be effective, but only if they include information about a way to avoid the source of the fear. Some Nazi propaganda, for example, focused on themes of Jews seducing Gentile women and cheating families out of their life savings, in order to instill a fear of Jews in Germans. To make the propaganda effective, however, the Nazis then listed local Jewish-owned businesses to avoid as a way to alleviate the fears they had generated (Myers 1996: 282–3).

Propaganda themes should reinforce existing beliefs, instead of attempting to instill new beliefs in a target audience. For example, even though the Soviets controlled all media in Poland, their attempts from 1945 until the end of the Cold War to make the Poles pro-Russian largely failed. This case illustrates how difficult it is to change basic beliefs, such as the Poles' distrust of the Russians, as opposed to reinforcing beliefs that already, at least to a minimal extent, exist in an audience even when the government controls all the means of mass communication (Rakowska-Harmstone 1990: 92–112). A similar phenomenon occurred during the Second World War, when the Allies learned that propaganda attacks on Nazi ideology had little effect. By the early 1940s, Nazism was well entrenched in Germany and it was extremely difficult, if not impossible, to shake most of the population's belief in it (Riley and Cottrell 1958: 542).

The groups that surround individuals and the norms they embody are powerful barriers to individuals changing their beliefs. In fact, people usually accept or reject persuasive messages based on whether the message is seen as diverging from the norms of their reference group or opinion leaders, who function as gatekeepers by promoting or rejecting beliefs and judgments, especially those transmitted by the mass media (Janis 1976: 609–19; Myers 1996: 291). Therefore, in a process called social validation, individuals are more likely to comply with a request for behavior if people similar to the individual appear to be thinking or doing similar things. After they have been persuaded, individuals are also much less likely to change their minds again if they endorse or are made to endorse their new position publicly. The Communists relied on this concept of consistency when they forced citizens, especially those undergoing re-education, to publicly state their support for the Communist regime (Cialdini 1995: 262–7).

Given the difficulty of changing beliefs, some of the most effective propaganda reinforces simple, basic themes that already exist in an audience, regardless of their political beliefs. For example, during the Vietnam War, a US psyops officer talking to a VC defector about possible propaganda themes found the ex-VC boiling down propaganda to one essential theme. He said, "you must understand that here in Vietnam people do not like to be killed" (Berger 1976a: 673). Such commonly shared basic beliefs can be found in all people and can be reinforced through psyops to override the will to fight for or to support any particular political movement.

Even if an effective theme has been selected, translation of propaganda into the target nation's language must suit the target audience. For example, during the Second World War, the United States dropped leaflets on Japanese cities in the form of fake extra editions of Japanese newspapers. The newspapers were written in a literary form called "bongobun," which was the style of Japanese newspapers up to 1930. Unfortunately, most of the Japanese living in the United States who were recruited into US propaganda had left Japan before 1930. After 1930, however, Japanese newspapers switched to "kogobun" or an oral form. With many old, out-of-style words in the newspapers, the Japanese easily identified the newspapers as having been written by foreigners, making them worthless as propaganda. Similarly, US soldiers laughed when early German propaganda leaflets in the Second World War called them "doughboys," referring to the First World War name for their forbears (Herz 1976: 678–9). The VC also often made awkward translation errors. One VC/North Vietnamese Army (NVA) leaflet quoted a US sergeant, "I would like to take this occasion to thank the Vietnamese people and the SVN-NLF for returning my freedom, once again.... I thank you for the Front's lenient and humane policy towards foreign ralliedmen and POWs" (Herz 1982: 352).

Translating often faces the problem that terms have different meanings in different places. In the Far East during the 1960s, for example, the United States often used the term "special warfare" to describe various military operations. To Far Eastern audiences, however, "special" meant secret police, bribery and assassination. Counterintuitively, keeping terms accurate can be even more difficult if two nations share the same language, because the same words sometimes have different meanings. For example, in India and Pakistan, states with a heritage of British rule, the word socialism is understood in the British sense of the term and means social welfare with some public ownership. The United States, by that definition, is very socialistic, yet Indians and Pakistanis hear US leaders deny that they are socialist. This only leads to confusion at best and charges of hypocrisy at worst (Linebarger 1982: 339–41).

Mistranslation, however, also can be done on purpose. In the First World War, for example, the British deliberately mistranslated a story from France where, they claimed, the Germans had shot a young boy for guerilla activities, when in fact the Germans had executed a full-grown man (Ponsonby 1928: 94–6). The Germans were not above mistranslations of their own. During the naval blockade of Germany, the Germans said that the diseases from which German children suffered were called *Die englische Krankheit*, reflecting British inhumanity. Long before the war, however, the phrase had been the common German name for rickets (Ponsonby 1928: 20).

Just as the text used in propaganda must be right for the intended recipients, images used in propaganda must be credible in the eyes of the target audience. In 1951 during the Korean War, the United States produced a poster showing a drawing of Chinese soldiers running from UN jets, tanks and artillery with the caption, "Death – in Many Forms – Awaits You on this Foreign Soil" (capitalization in original; Daugherty 1958h: 751–3). A panel of cooperative Chinese POWs evaluated the poster. Not one of the Chinese POWs interviewed saw the image of the Chinese soldiers as being Chinese. The drawing, it turned out, was what

Americans thought a Chinese soldier looked like. In fact, many of the Chinese POWs concluded the drawings must be of American soldiers.

Propaganda can be distributed through all means of human communication, including, but far from limited to, slogans, speeches, songs, banners, paintings, statues, motion pictures, stamps, postmarks, e-mails, letters, websites, rumors and photographs. In modern times, however, the press, radio, television and the Internet are the most common methods for distributing propaganda. The key is to select the proper medium to reach the target audience. The VC, for example, given the lack of radios, let alone televisions in rural areas of Vietnam, mainly relied on lower-tech methods, such as slogans, songs, speeches and posters to disseminate propaganda (Gessner 1976: 680–2). By the 1970s and 1980s, when radios had become common in the Third World, rebels in El Salvador, Mozambique, Cambodia and Nicaragua turned enthusiastically to radio, even establishing their own radio stations (Radu 1990: 118–43).

As the following examples suggest, propaganda can encompass a vast range of methods of dissemination. At the high-tech end of the spectrum, direct broadcasting satellites (DBS) allow states to transmit television shows into another state without seeking the other government's permission. The United States, for example, planned to insert its own propaganda into Haitian television broadcasts when it considered intervening in 1995, but did not do so when the invasion was canceled. The cost for such an operation is relatively low. A DBS transponder over Asia, for example, can be leased for about $2 million per year, and will probably decrease in cost in the future, while an entire system can be launched for about a billion dollars: well within the reach of many nations (Libicki 2000: 46).

At the low-tech end of the spectrum, however, little technology is required to spread propaganda. The Psychological Warfare Department of the Defense Ministry in Taipei, Taiwan, for example, sent 101,614,528 balloons carrying leaflets floating serenely toward mainland China between 1959 and 1976 (McLaurin *et al.* 1976c: 727–8). During the Second World War, the British also used various relatively simple media to deliver propaganda. The British Morale Operations (MO) branch distributed phony German newspapers in Germany that praised Hitler to add to their authenticity, but also intimated that the Third Reich was doomed. In another operation, Office of Strategic Services (OSS) agents, the forerunner of the CIA, planted fake newspaper clippings on dead German soldiers, then slit their wrists. If Germans found the bodies, they were meant to assume the soldiers had committed suicide after reading the newspaper, which included a story about an order from SS chief Heinrich Himmler that any German wife of child-bearing age who had not had a baby during the past two years would have to report for duty to a SS "breeding farm." Not the best news for a married soldier's morale (Breuer 1988: 268–9).

The mails also can be used for propaganda. During the Second World War, the British MO branch mailed fake "official" death notices to families of German soldiers fighting at the front, while near the end of the war, the United States launched Operation Cornflakes. The goal was to deliver propaganda through the mail at breakfast, hence the name "Cornflakes." To get the "mail" delivered, the United States bombed mail trains bound for Germany from Italy. Mailbags containing propaganda letters, which had been carefully created to appear authentic, were then dropped over the site. The hope was that the Germans would assume they were real mail bags, since the Americans went to great trouble interviewing POWs and people who had lived in Germany to learn about postal system markings and procedures, and deliver the letters as normal mail. Some letters were from the League of German Partisans and suggested a widespread peace movement in the army. Other letters appeared to be from "The League of Lonely Women" and were addressed to combat troops. The letters said;

> When are you coming on leave? ... We are waiting for you in any strange town you may pass through. Cut off the League symbol from the letter. Stick it on your glass when you come in any café, in any bar near a station. Soon a member will be with you, and all the traumas of fighting will disappear in the beauty of one night.... Don't be shy, your wife, sister or loved one is also one of us.
>
> (Moynahan 1976: 763–5)

Even postal marks can be used for propaganda. In March 1957, Hungary protested to the United States and Canada about cancellation marks on letters entering Hungary that said, "Support your Crusade for Freedom" and other slogans supporting reform in Eastern Europe. The Soviet and Hungarian Communist press covered the story extensively, violating the first rule of propaganda: do not argue your opponent's case in public (Choukas 1976: 726–7).

Even an encyclopedia can be used for propaganda. The British IRD used entries in the Soviet Union's own *Soviet Encyclopedia* to highlight Soviet deportations, forced labor practices and the disappearance of entire ethnic groups. They supplied stories to the media, for example, that certain ethnic groups had disappeared from the 1950 encyclopedia, although they had been included in earlier editions (Aldrich 2002: 456). The IRD, however, did not stop at analyzing other peoples' books. They published their own, following in the footsteps of earlier propagandists, including Thomas Paine (*Common Sense*), Adam Smith (*The Wealth of Nations*) and Karl Marx (*Das Kapital*). In 1950 the IRD published Robert Carew-Hunt's *Theory and Practice of Communism*. It had been written as an in-house IRD guide, but the British government cleared it for publication and Penguin published several printings. More IRD books followed in the 1950s and 1960s, including Robert Conquest's *The Great Terror: Stalin's Purge of the Thirties*, which drew heavily from IRD files. Most of the books were written by freelancers and published by front companies, such as Ampersand Books, which had been set up by the IRD's Leslie Sheridan. Ampersand published 20 books over three decades, but lacked a large distribution network, so the IRD teamed with The Bodley Head Press to distribute the Background Books series, which was a huge success. The books were written by intelligence men, such as Robert Bruce

Lockhart and Monty Woodhouse. Bodley Head then started publishing memoirs of intelligence operatives, blurring the line between publishing for profit and for propaganda (Aldrich 2002: 458).

Propaganda images can also take the form of comic books, which allow propagandists to reach even the illiterate. During the March 1982 election campaign in El Salvador, for example, the CIA distributed color comic books. In an attempt to achieve a centrist victory, some showed leftist rebels torturing children and small animals, while others were anti-right wing, showing rich, fat men in suits leaning against their limousines while surrounded by armed thugs (Tayacan 1985: 23).

The type of medium, whether DBS-transmitted television program or balloon-delivered pamphlet, influences the effectiveness of propaganda. In general, the more lifelike the medium, the more persuasive the message. Studies have shown that the order of persuasiveness of types of propaganda is live, videotaped, audio-taped, photographed and then written forms (Myers 1996: 283, 291, 294). Because of this, altered images on film, videotape and in photographs could, and sometimes have played a significant and effective role in psyops. In fact, images can be so powerful that they distract an audience from accompanying text or audio. For example, the famous 1964 anti-Goldwater political advertisement created by adman Tony Schwartz showing a little girl in a field of daisies that transitioned into a mushroom cloud, and images of 1988 Democratic presidential nominee Michael Dukakis encumbered by a helmet and riding in a tank were remembered long after the words and issues of the campaigns were forgotten.

Whatever the medium, the audience must pay attention to the propaganda for it to have an effect. It would seem that getting a propaganda message noticed would be relatively straightforward, yet great difficulty lurks in even seemingly simple things. For example, as the Allies were fighting their way out of the Normandy beachhead, they dropped leaflets on French villages warning of impending bombings. The French frequently ignored the warnings and a psyops team was assigned to find out why. In interviews with French villagers in newly liberated areas, the villagers said they had seen the leaflets and had understood the message, but had not thought that the leaflets were meant for them. They assumed the leaflets must have blown over from some other town, since no one would have any reason to bomb their little village (Riley and Cottrell 1958: 537).

The labeling of a message can have a significant influence on whether the target audience notices it. Usually this involves making an image vivid and eye-catching or using a good headline for a story, but it can also be as simple as a label. Crossman, the British propagandist, said, "If you want to get a rumor read by important generals, admirals, and air marshals, there is nothing like getting it included in a report marked 'secret'" (Daugherty 1958a: 37).

Propaganda can target civilians, government officials or the military in enemy, neutral or allied states. No matter what the target audience, for propaganda to be effective, the propagandist must know the target audience intimately. The Soviets, for example, learned that all nations could not be fooled in the same way.

"German journalists tended to believe forged documents as long as they looked official. British journalists needed to believe the source... Indians and Egyptians would accept practically anything so long as it supported some kind of conspiracy" (Naftali 1999: 7).

Psyops, however, usually do not target an entire population because propaganda will never change the mind of a true-believer. Reception analysis has shown how different segments of an audience can interpret propaganda in different ways; often ways the propagandist did not intend (Berger 1995: 110–11; Underwood 2003: Online). Therefore, propaganda is usually targeted at the most susceptible segments of a population. Henry V. Dicks, for example, a psyops researcher during the Second World War, established five categories of German males of military age: fanatical "hardcore" Nazis (10 percent), modified Nazis with reservations (25 percent), unpolitical Germans (40 percent), passive anti-Nazis (15 percent) and active anti-Nazis (10 percent). Therefore, most German soldiers were politically neutral, a fact of critical importance to Allied officers designing propaganda (Katz 1976: 482). Dicks' research highlights the requirement for extensive intelligence work about the attitudes of a target population. During wartime, information can be gathered from POWs, captured documents, intercepted mail and by monitoring enemy newspapers and broadcasts, as well as by interviewing academics, businessmen and ex-patriots who know the target state.

In at least one case, however, extensive intelligence research led to trouble, not because it was poorly done, but because it was too good. In the early 1930s, Bethold Jacob published a book on the German military that detailed the new General Staff, army group commands and districts, and even listed the rifle platoons attached to the new panzer divisions. He also provided thorough biographies for 168 generals. Hitler was furious and ordered an investigation into this serious breach of military security. On 20 March 1935 in Switzerland, German agents invited Jacob to dinner, then drugged and kidnapped him. Under interrogation back in Germany, however, Jacob explained that all of his research had come from German newspapers. He had diligently combed through the society pages, obituaries and news articles to find the names of officers and units, including where they were based, their units' strengths and compositions. When an officer married, for example, often the entire staff of his unit would be present at the wedding and were listed in the newspaper. Officers' obituaries also provided a treasure trove of information on various units in describing the military career of the deceased. Luckily for Jacob, his interrogators, who admired good intelligence work, believed him. He was later released (Farago 1958a: 514–16).

The vast literature on the effects of propaganda has found significant variations in those effects. In the 1930s and 1940s, most theorists believed that propaganda had significant effects, which led to the bullet, transmission belt or hypodermic needle model: whatever message was communicated led to significant changes in attitude in the target audience. In the 1950s and 1960s, a limited-effects model was developed: the mass media were but one of many factors affecting beliefs and opinions, with opinion leaders severely limiting attitude change. By the 1970s, the evidence appeared to show that propaganda could have significant effects on certain narrow topics (Severin and Tankard 2001: 125–6, 268–82). More recently, Berger (1995: 73) argued that there has been little evidence found of the effects people believe are caused by mass communications (Baran and Davis 1999: 263–5), although there does appear to be a socialization effect (Berger 1995: 62–3) and a significant cultural impact on other nations (Berger 1995: 71). Disagreeing with Berger, Foss (2004: 310) argued that images in the mass media have significant effects, although these effects are much harder to define than the effects of words. Much of the difficulty lies in clearly defining the message in an image to then determine if an attitude change occurred in response to that image.

For all the changes in theories about the effects of propaganda, most researchers believe that propaganda has effects, although it has been difficult to measure those effects. Even so, it is safe to rule out the possibility that propaganda is completely ineffective. More importantly for the focus of this book, states and groups will continue to use propaganda because many leaders believe that propaganda has a significant effect.

The commonly made argument that people today are better able to detect propaganda than their predecessors because they can easily identify propaganda from the past is an invalid argument. No matter how educated, democratic or media-savvy a population becomes, people will never be immune to propaganda that is created by a skilled propagandist. Effective propagandists analyze a population, determine what that population believes, and then devise messages that reinforce the parts of existing beliefs that the propagandist wants to bring to the fore. This trait makes propaganda specific to time and place. Therefore, propaganda from different time periods or even from other nations often seems heavy-handed and coarse because the attitudes that the propaganda was designed to reinforce do not exist in populations outside the target audience. For example, in the 1930s, *Triumph of the Will* reinforced beliefs that most Germans already held: that a strong and united Germany should arise from the defeat in 1918. Because it supported these commonly held beliefs, many, if not most, Germans loved the film. Non-Germans, who did not hold these beliefs, dismissed the film as blatant propaganda and discounted its impact on Germans (Rohatyn 1988: 79–80). The difference in the effect depended on whether the audience already held the beliefs the film reinforced. This does not mean, however, that propaganda does not evolve as people become attuned to its techniques and new methods are devised. Propaganda must, to a great degree, be covert to succeed, although the labeling of a message as propaganda often has more to do with the audience disagreeing with the message than with how overt or covert the techniques of the propagandist.

Apart from the debate over whether propaganda has a significant effect, psyops and propaganda do have certain widely acknowledged limitations and weaknesses. Psyops take extensive planning and deployment time. There is often a lack of qualified personnel, such as linguists and experts on a target state to plan and develop psyops material. Furthermore, unintended audiences, especially in friendly or neutral states, may also "hear" the propaganda, causing harmful unintended effects. As a final hindrance to psyops and propaganda operations, moral considerations, domestic politics and laws restricting their use can impose limitations, especially in democracies.

The most important limitation of psyops, however, is that they cannot be effectively conducted alone as a substitute for military operations. Although they are often organized as a separate sub-branch of the military, psyops cannot operate in isolation. Crossman said, recalling his service as head of SHAEF propaganda during the Second World War, "Unfortunately, soldiers, sailors, and airmen usually resort to psychological warfare when they are in a fix and can think of nothing else to do" (Daugherty 1958a: 37).

> When the soldier came to us and said, "I cannot solve my problem by military means; would you please provide me with a psychological warfare substitute?" we always replied, "No; I cannot, I will not, I must not." *Psychological warfare is no substitute for action.*
>
> (Italics in original. Daugherty 1958a: 44)

Psyops always must be coordinated with military, economic and diplomatic policy or they may do more harm than good. In 1941, for example, the BBC started the V for Victory campaign, which led many in occupied Europe to believe that an invasion of Western Europe was imminent. This was not the case, but the campaign led many Europeans to mount attacks against the Germans prematurely that resulted in fruitless Resistance losses. The 1943 Allied propaganda campaign about the opening of a second front, which was designed to deceive the Germans into keeping more forces in the West, even though an invasion was not planned for that year, also led to many Resistance casualties. The lack of coordination between psyops and military plans also meant that the Germans kept some units in the west that otherwise might have been moved to other fronts. These units were later in place to defend against the 1944 Normandy landings, highlighting the requirement to coordinate psyops with military, economic and diplomatic policy (Daugherty 1958a: 36).

4

PSYCHOLOGICAL OPERATIONS

The un-American weapon?

Even though the United States has used psyops throughout its history, the American public generally views psyops and propaganda as dirty, underhanded and unfair. There was even a book on propaganda called *The "Un-American" Weapon* (Harris 1967). This distaste for psyops is ironic, given that during the American Revolution many colonists had grave doubts about breaking with Britain, so the revolutionaries routinely used propaganda to further their cause. The Declaration of Independence itself was mostly propaganda that sought to justify the revolution. Samuel Adams even formed the Committee of Correspondence to gather news, modify it with a pro-revolutionary spin, and distribute it to newspapers.

One of the most famous examples of colonial propaganda was based on an incident in March 1770, when 20 British soldiers faced a mob of street toughs in a labor dispute in Boston. As discipline broke down, the soldiers opened fire, killing three colonists outright, while two more later died of their wounds. Calling the incident the "Boston Massacre," the revolutionaries made certain that the news, suitably exaggerated to harm the British, was spread far and wide. Recognizing the power of images to mold public opinion, Paul Revere engraved a scene of the "massacre" that showed disciplined British troops cold-bloodedly shooting peaceful, respectable citizens. Revere's version increased the number of dead and even changed the race of one man who had been killed, Crispus Attucks, from black to white. Although a colonial jury later found the soldiers not guilty, the "massacre" was a significant boon to the revolutionary cause.

Thomas Jefferson and Benjamin Franklin served on a committee that sought to persuade Hessians and other foreign troops serving in the British Army to defect (Butterfield 1958: 66). One of the committee's tactics was to have George Washington's men on Prospect Hill throw notes wrapped around rocks to the British as they faced each other outside Boston. The notes stated that on the colonists' hill there were "seven dollars a month," "fresh provisions and plenty," and "health, freedom, ease, affluence and a good farm." By contrast, on British-held Bunker Hill, there were "three pence a day," "rotten salt pork," "The Scurvy" and "slavery, beggary and want" (Daugherty and Janowitz 1958: 60).

During the Civil War, both the North and the South employed psyops and propaganda. The Emancipation Proclamation, for example, was a powerful piece of psychological warfare. Although the Confederacy was the biggest supplier of cotton to British textile mills, Abraham Lincoln hoped the Proclamation would play on British anti-slavery sentiments and ensure that London did not support, let alone recognize the Confederacy. By attempting to turn the war into a fight over the issue of slavery, Lincoln also hoped to convince most Southerners, who did not own slaves, that they were fighting a war for rich, slave owners. In contrast, Jefferson Davis and the Confederate leadership produced propaganda that argued that the South was fighting not for slavery, but for states' rights, which had broad support in the Confederate states (Janowitz 1958a: 73–9).

In August 1917, the United States established the Committee on Public Information to coordinate propaganda for the First World War, although it became better known as the Creel Committee after its chairman, George Creel. To ensure the close coordination of psyops with military and diplomatic policy, the secretaries of state, war and the navy served on the committee. The committee arranged voluntary censorship of all news and managed extensive propaganda campaigns that included stickers, buttons, posters, books and films. The committee even leaked rumors favorable to the war effort to the media. They also trained teachers and speakers, known as "Four-Minute Men," who spoke in schools and at civic clubs across the country in support of the war. President Woodrow Wilson's Fourteen Points speech was also a powerful propaganda weapon. Because the Fourteen Points made the terms for peace appear more appealing than continuing the war, the Germans and Austro-Hungarians had difficulty countering its allure after the United States and its Entente allies directly targeted citizens of the Central Powers. The United States distributed more than a million copies of Wilson's speech in Germany alone (Sisson 1958: 88–92).

After victory in 1918, US psyops almost ceased to exist until the rise of the Axis in the 1930s, when Washington established the Institute for Propaganda Analysis to monitor Nazi themes in the media. During the Second World War, the Office of War Information and SHAEF's Psychological Warfare Division developed psyops to new heights, producing newspapers, newsreels, radio broadcasts, motion pictures, posters, songs and all manner of propaganda for American, allied and enemy audiences. One tiny facet of this momentous psyops operation included the dropping of millions of leaflets over Japan that warned of impending fire bombing raids. This effort climaxed during the closing months of the war; on a single day, 30 July 1945, the United States dropped 500,000 leaflets. These forewarnings increased the impact of bombing attacks and reduced civilian resentment. The leaflets emphasized that a given city was being bombed because it was a military target with factories and military installations, "But, unfortunately, bombs have no eyes." Therefore, the leaflets urged, civilians should leave the city. Besides being a humanitarian gesture, if the populace fled, it disrupted the Japanese war effort because fewer workers showed up for their jobs at factories (Daugherty 1958c: 360). The propaganda leaflets could even have a powerful effect if there was no follow-up bombing. In 1945, a single US plane dropped leaflets on a railway marshaling yard in China, warning the Chinese rail workers to flee because the yard was about to be bombed. Although the bombing failed to materialize, the workers fled and it took 10 days for the Japanese to round up enough workers to get the railway operating again (Daugherty 1958c: 362).

After 1945, US psyops again went through a precipitous decline, although in 1949 C.D. Jackson, publisher of *Fortune* and Eisenhower's top wartime advisor, established Radio Free Europe and Radio Liberty to transmit propaganda into communist Eastern Europe and the Soviet Union. On 20 June 1950, President Harry Truman established the Psychological Strategy Board and in 1952 the United States founded a Psychological Warfare Center at Fort Bragg, North Carolina. The Americans and their British allies also attempted to use propaganda to incite purges in Eastern Europe in the late 1940s and early 1950s by playing upon Stalin's paranoia. Western propaganda linked opposition movements to top Eastern European Communist officials. The operations did trigger purges, but backfired when they failed to spark disorder and a weakening of Soviet rule, but instead led only to a consolidation of Soviet control (Aldrich 2002: 178–9).

The United States also used propaganda in the form of short films to promote the Marshall Plan in Western Europe from 1948 to 1953. These were subtle by design. For example, they only mentioned the plan by name twice in one-reel films and three times in two-reelers, focusing instead on the benefits of economic growth and cooperation. Few Americans saw the films since unlike some politicians today, Congress then believed that it was wrong to use US taxpayers' money to propagandize in the United States, so it was illegal to show the films in the United States until 1990 (Turan 2005: E1, E16).

In a more focused effort, an extensive US propaganda campaign helped deal the Communist-led Popular Front a defeat in the 1948 Italian elections. The Italian Communist Party was one of the strongest in Western Europe and was expected to garner at least 40 percent of the vote, but received only 31 percent. One of the US propaganda campaign's most effective weapons was a 1939 MGM motion picture, *Ninotchka*, starring Greta Garbo and Melvyn Douglas, that satirized life in the Soviet Union. After the election, a pro-Communist worker said, "What licked us was Ninotchka" (Janowitz and Marvick 1958a: 322). Each week, US-produced documentaries were shown to more than five million Italians, while another eight million watched pro-Western, Italian-made documentaries. The United States also organized a massive form-letter campaign. The letters were distributed to Americans of Italian descent, who then filled in the name and address of relatives and friends in Italy. Hundreds of thousands of these personalized letters were sent, warning that a Communist victory would lead to the loss of American aid, the decline of the Catholic church, and a less prosperous and independent Italy (Janowitz and Marvick 1958a: 320–6).

During the 1950s US propaganda was also aimed at allies, such as Britain. American propaganda focused on schools, the young and labor unions as possible bulwarks against Communists. The Americans founded a liberal think tank, Socialist Union, and a monthly journal, *Socialist Commentary* that supported Anglo-American cooperation. They also published pamphlets, such as

British Agents of the Cominform, to sell through British newsstands, and in 1953 published a pro-American academic magazine, *Encounter*, similar to *Foreign Affairs*. Although the US backing of *Encounter* was obscured, *The Times Literary Supplement* detected an obsession with the evils of communism and T.S. Eliot guessed correctly that it was US "propaganda hidden under a veneer of British culture" (Aldrich 2002: 447–50). The US also tried to send British labor leaders to the United States on cultural exchanges to try to make them more sympathetic to America. Ironically, US immigration blocked some of the individuals from visiting because they were former Communist Party members. To combat the growing anti-Americanism in Britain that accompanied the rise of McCarthyism during the early 1950s, US propaganda efforts overseas expanded significantly. By March 1953, for example, there was a US propaganda staff of 93 operating in London with an annual budget of $850,000 (Aldrich 2002: 452).

As had happened during the world wars, US psyops expanded greatly during the Korean War. At the height of the conflict, the United States produced 200 million propaganda leaflets a week. Some of the leaflets warned of imminent UN bombing. The leaflets increased UN credibility because they were accurate and undermined enemy confidence because they gave the impression that UN forces were so strong that the North Koreans could not prevent bombing attacks even if they were forewarned (Harris 1976: 412). Other leaflets resorted to bribery. A Harvard study concluded that while Westerners viewed bribery as immoral, Asians perceived bribery as a normal way of doing business. In Operation Moolah, a leaflet offered $100,000 to the first North Korean pilot to defect with a MiG fighter, and $50,000 for each subsequent pilot. Although no pilot appears to have come south for the money, the leaflets still had a significant effect. After the leaflets appeared on 16 April 1953, the North Koreans grounded their MiGs for eight days. Even after the jets resumed flying, there were far fewer in the air, which appeared to have a significant effect on the air-war. During the 60 days before 16 April, the United States shot down 53 MiGs for the loss of 4 Sabers, whereas during the 60 days after 16 April, US pilots shot down 107 MiGs for the loss of only one US plane. The leaflet also had repercussions throughout the lands behind the Iron Curtain. From then on, Communist block pilots underwent far more political indoctrination and spent far less time honing their air-combat skills (Berger 1976b: 689; Murray 1958: 417–18).

Even as psyops enjoyed a resurgence in the 1950s, the United States made several serious missteps, such as during the 1956 Hungarian uprising when psyops were not coordinated with US diplomatic and military policies. After the Soviets crushed the uprising, there were charges that the Voice of America and Radio Free Europe had incited the rebellion, only to abandon it. Studies by West Germany and the Council of Europe failed to find any explicit or even implicit promises of US support in broadcasts at the time, yet many Hungarians believed such promises had been made. Unfortunately, US policy had been ambiguous during the uprising and Washington only publicly disavowed the use of force after the rebellion was over. American credibility was gravely harmed by this sin of omission (McLaurin, Rosenthal and Shillings 1976a: 382–5).

The Vietnam War posed special problems for US propagandists. It had been far easier to rally support for democratic Britain in a total war against the Axis powers where unconditional surrender was the goal than it was to seek support for a far from democratic South Vietnam in a struggle to maintain the pre-war status quo of a Communist North Vietnam and a Western-leaning South Vietnam. Worse, some of the lessons of the Second World War had been forgotten by the time of Vietnam. One poster used in Vietnam, for example, showed a US rural development team repairing a road without a single villager in sight. The poster missed an opportunity to convey Vietnamese–American cooperation, which would have been the message if villagers had been shown working with the Americans. Some US posters aimed at the VC also portrayed VC soldiers in a negative light, which was unlikely to convince them to surrender. Other US psyops, however, were remarkably effective, especially on the battlefield. In September 1966 in the First Infantry Division's area of operations, a guerrilla surrendered. The Big Red One's intelligence staff made a tape of the man calling to his former comrades by name, which made the appeal extremely creditable. Within 24 hours, 88 more guerrillas defected (Pappas 1976: 277–9).

Even with such tactical successes, the US emphasis in Vietnam was on high technology and firepower, not ideas. The United States only had 1,200 American and 750 Vietnamese working in psyops at the height of the war, and the Joint US Public Affairs Office had a budget of only $12 million at a time when the United States had 500,000 men and was spending billions of dollars a year in Vietnam. Although even this limited effort led by 1967 to the United States dropping 5 billion leaflets in Vietnam, operating a four-station radio network that reached 95 percent of the population and a television station that broadcast six hours of programming a day (Shultz 1989: 123–4), the United States lost the psychological war both in Vietnam and America.

As with other wars, after the Vietnam War there was another decline in US psyops. In the mid-1970s, US black propaganda ceased altogether after Congressional investigations of the CIA severely criticized such activities. Psyops were largely comatose until the development of information warfare (IW) in the late 1980s, which prescribed using information, including psyops and propaganda, to win wars. During the Gulf War, for example, US propaganda leaflets dropped on Iraqi positions emphasized the devastating effect of precision Coalition air attacks, which convinced many Iraqi soldiers, and sometimes entire units, to surrender (Hall 2003: 114).

During the 1994 Haitian operation, the United States used an EC-130 aircraft carrying two 10kW transmitters, called Radio Democracy, to broadcast pro-American and pro-Aristide propaganda to the island. On the night of 15 September, the Pentagon parachuted thousands of radios, pre-tuned to the correct frequency, to ensure that the impoverished Haitians would be able to hear the messages. Leaflets reading, "The Sun of democracy/The Light of Justice/The Warmth of Reconciliation/With the Return of President Aristide," were also dropped over Haiti. US IW experts even sent derisive e-mails to top officers in the Haitian junta to undermine their morale. In the final stages of the crisis, when President Carter went to Haiti to negotiate, CNN was tipped off so that it was able to cover the liftoff of the 82nd Airborne division from Pope Air Force Base in North Carolina on its way to invade Haiti. The images may possibly have given an added impetus to the Junta to leave Haiti and go into exile (Adams 1998: 84–5).

A similar case of the United States and the British using the media for psyops occurred in the 1990s when Libya's Qaddafi sent agents to meet with a top South African biologic weapon developer to buy his skills and research. Western intelligence agencies had penetrated Libya's procurement program, however, and knew of the meeting before it took place. In February 1995, a US or British agent leaked the story to the press. The exposure worked. The South Africans destroyed their biologic weapons research and Libya's attempt to acquire such weapons was postponed (Adams 1998: 275–6).

The United States also used propaganda during and after the 2003 Anglo-American military intervention in Iraq. For example, after the invasion the US planted stories in Iraqi newspapers that were basically factual, but omitted any criticism of the US or Iraqi governments (Mazetti and Daragahi 2005: A1, A12). In another example, early in the conflict, the rescue of a POW, Private Jessica Lynch turned her into a hero. Early *Washington Post* reports focused on Lynch's heroics in fighting Iraqis when her maintenance unit was ambushed. She only surrendered, it was reported, after suffering multiple bullet and knife wounds. None of it, however, was true. It is unclear where the story originated, especially since Lynch, showing, as columnist Scheer (2003: E4) argued, great honor and impressive fortitude, refused to join the media bandwagon of her alleged heroics. Whatever the source, the story of the ambush of her unit and her subsequent rescue are clearly examples of propaganda. Given that the US military filmed her rescue, albeit from what turned out to be an undefended hospital, one wonders if forthcoming US military operations will have producers, directors and cinematographers attached to ensure maximum propaganda value from each mission. If so, such coverage will only continue a tradition of US psyops and propaganda going back to Paul Revere's engraving of the Boston Massacre and may, if done well, be just as effective.

The United States, of course, is far from the only nation to use psyops and propaganda. Although images have not played a significant role in most psyops, in part because of photography's relatively recent development, a brief overview of some past psyops will provide a sense of the range of operations that have been undertaken and their general approaches. Almost all of these operations could be copied in the future using altered images.

The Vikings and Mongols relied on stories of their martial exploits and barbarism spreading before their advances to intimidate would-be foes and ease their conquests. The Romans, who also understood the power of psychological effects, created the term Carthaginian peace when they obliterated their rival Carthage. The Romans killed or enslaved the entire population, leveled the city and salted the soil so nothing would grow, in part, to send the clear message to the rest of the world: do not oppose Rome. In modern times, the Mexicans focused propaganda on recent Irish immigrants in the US Army during the 1846–8 Mexican–American War. The Irish were Catholic and the Mexicans exploited their shared religion to convince some Irish-Americans to desert or, in some cases, to switch sides (Andrews 1958: 72–3).

With the rise of nationalism and mass media with its ability to distribute propaganda rapidly and widely, the twentieth century saw a concurrent rise in psyops and, especially, propaganda. In 1899, for example, Japan hired a German to produce pro-Japanese stories for the Western press to help foster more positive views of Japan (Mercado 2005: 48). In 1903, Tsar Nicholas II's "Okhranka" or secret police drafted a document that purported to be an agreement between Jewish elders meeting in Switzerland in 1897 to plot the destruction of Christian civilization and its replacement by a Jewish hegemony. *The Protocols of the Learned Elders of Zion* was published in Russia on the eve of a 1903 pogrom in Odessa to justify attacks on Jews. The document later became the "most influential forgery of the twentieth century" (Andrew and Gordievsky 1990: 25–6) when the Nazis and Hitler, specifically in *Mein Kampf*, cited the document to justify their anti-Semitic policies. The document is still being cited in the Middle East. Egyptian television in 2002, for example, produced a 41-part series based on *The Protocols* (Boot 2005: B13).

During the First World War, all types of propaganda flourished. Even official documents were falsified, including casualty lists in almost every country. Lists were shortened and differed more and more from reality with each month's mounting casualties (Ponsonby 1928: 140–5, 167–75). British politician Stanley Baldwin concluded, "In the arena of international rivalry and conflict men have placed patriotism above truthfulness as the indispensable virtue of statesmen" (Ponsonby 1928: 11). Exaggerated stories were common. For example, on 29 August 1914, the official British Press Bureau reported, "The intellectual metropolis of the Low Countries (Louvain) is now no more than a heap of ashes," while *The Times* reported that "Louvain has ceased to exist," even though only about one-eighth of the town had been damaged, let alone destroyed (Ponsonby 1928: 21). British newspapers also published extracts of letters from the front describing glorious, albeit fictional victories, often written by bored soldiers to impress lovers back home (Ponsonby 1928: 23). There were also countless spy stories. British stories often had a governess, waiter or tradesman leading a double life as a German spy (Ponsonby 1928: 152). Oddly, the British rarely, if ever, had any Austro-Hungarian spies in their stories.

With government encouragement, the media also ensured that nothing to the enemy's credit was disseminated. A British war correspondent who mentioned a German's chivalrous act toward an Englishman, for example, received a telegram from his editor: "Don't want to hear about any good Germans" (Ponsonby 1928: 21). The images of the enemy, however, changed during the war. At first, the British and French portrayed the Germans as cowards and useless soldiers whose greatest desire was to surrender. As the war continued, the British and French people began to wonder; if the Germans are so cowardly and useless, why aren't our troops in Berlin yet and the war over? Stories about the cowardly Germans soon vanished from British and French propaganda (Ponsonby 1928: 24).

Atrocity stories with little or no foundation in truth were spread in magazines and newspapers. As John Bright, an English politician, said, "You will find wars are supported by a class of argument, which, after the war is over, the people find were arguments they should never have listened to" (Ponsonby 1928: 11). British atrocity stories included, among many others, U-boat crews shooting women and children in lifeboats, German soldiers mutilating nurses, ripping the tongues out of British POWs, chopping the hands off a Belgian baby, crucifying a Canadian officer, bombing hospitals and instituting germ warfare (Ponsonby 1928: 67–70, 78–82, 91–3, 97–8, 116–18, 154–6). Truth was indistinguishable from lies, so soon most of the public believed little, if any, news, even if it was true. Propaganda exaggerated victories and minimized defeats, unlike during the Second World War, when the British, especially, sought credibility through telling the truth. The British policy of truth was, in part, a reaction to the lack of credibility propaganda had attained by the end of the First World War.

In the First World War, propaganda was even used to attempt to sway entire peoples. The 1917 British Balfour Declaration helped gain Jewish support for the Entente war effort (Steed 1958: 262–3). Both the Entente and the Central Powers also desperately wished to sway another people: the Americans. Both alliances were extremely active conducting psyops in the United States before her entry into the war. German methods were subtler at first as they sought to keep the Americans out of the conflict. A German-run news service provided war news that was accurate at first, but then was slowly slanted toward a pro-Central Powers' perspective. The British, in comparison, were much cruder, but far more successful. British propaganda portrayed the German soldiers as barbaric, bloodthirsty Huns and U-boat crews as modern pirates.

The warring states also came to realize the power of images to influence public opinion. At the start of the war, for example, only two British photographers and no artists were permitted at the front. Both photographers were Army officers assigned to create a historical record, not to provide photographs for the press or for propaganda. The penalty for anyone else foolhardy enough to take a picture in the British trenches was death. In 1916, however, Charles Masterman, head of British propaganda, decided he wanted illustrations for his publications. The government agreed and by 1918 more than 90 artists were in the British trenches supplying images for the press and Masterman's propaganda (Knightley 1975: 99). Although more popular in France than in Britain or Germany, the demand for atrocity photographs was especially great. A Vienna firm even offered atrocity photographs with blanks for the captions so they could be used by either side (Ponsonby 1928: 21).

After the war, many on both sides of the conflict concluded that propaganda had been a decisive factor in the Entente victory. There was even a book *Words That Won the War* (Mock and Larson 1939) about the decisive impact of propaganda on the outcome of the First World War. The German army did not appear to have been defeated since it marched home in good order after the armistice. The will, not the ability to fight appeared to have been the crucial variable, and propaganda was targeted at that variable. Therefore, after the war, propaganda was seen as a potent war-winning weapon. By the 1930s, propaganda had become a part of every war. When Italy invaded Abyssinia in 1935, Emperor Haile Selassie faced almost certain defeat on the battlefield, so he turned early to propaganda as a weapon. On the first day of the war he charged Italy with bombing a Red Cross camp. It is difficult to believe that even the Red Cross could construct a camp and fill it with refugees in less than a day so that the Italians could then bomb it (Knightley 1975: 187). Regardless of the veracity of the story, Selassie's propaganda worked in that the British, French and much of the world condemned Italy's invasion, although the condemnation and subsequent League of Nations embargo failed to prevent Italy from conquering Abyssinia. Selassie lost his throne, but the propaganda paid off in the long run because in 1941 the Western allies helped him regain his throne.

The Soviets used propaganda to maintain internal political control and as a tool of foreign policy. For example, in August 1933, the Soviet Comintern produced a book in London, *The Brown Book on the Hitler Terror and the Burning of the Reichstag*, that became one of their most successful pieces of propaganda. The Nazis charged that a Dutchman, Marinus van der Lubbe, had burned down the Reichstag on the night of 27 February 1933. He had been arrested half-naked and dazed at the scene and the Nazis charged that he had acted on behalf of Communists and Jews. The Nazis used the fire to justify arresting 4,000 people, mostly Jews, Communists and Nazi opponents. The incident was also used to justify the granting of emergency powers to Hitler. *The Brown Book*, however, claimed that the Nazis started the fire themselves to spark pogroms against Jews. This conspiracy story remained credible in many books into the 1970s, even though the Soviets completely fabricated it (Andrew and Gordievsky 1990: 187–9).

Drawing a lesson from what they believed had been the influence of propaganda on the outcome of the First World War, the Nazis used propaganda extensively. Propaganda was one of their first weapons and after they gained power it became, as in the Soviet Union and all totalitarian states, a tool of domestic social control. Hitler wrote, "The task of propaganda . . . is to blanket every area of human activity so that the environment of the individual is changed to absorb the movement's world-view" (Kumata and Schramm 1958: 50). The Nazis even established a college to train propagandists. Joseph Goebbels and his Ministry of Public Enlightenment and Propaganda ensured coordination of all German propaganda with access to all government planning and policy-making. "At times it seemed as if the whole German government operated as a subdivision of the ministry. Policies were carried out or discarded, depending on their propagandistic effect" (Kumata and Schramm 1958: 52). Goebbels even wanted control of courthouse reporters, so they would report only on verdicts that supported Nazi ideology. The Nazis also used images in their propaganda extensively, exemplified by the film, *Triumph of the Will*, and vivid posters with images of sharp-featured storm troopers, eagles and swastikas. After they gained control of the state, the Nazis acquired the weapons of armed force, the economy and diplomacy. Even with these other tools, the Nazis still relied on propaganda so much both domestically and as a tool of foreign policy that the Allies alleged at the Nuremberg trials Nazi propagandists "inject[ed] poison into the minds of millions and millions" (Myers 1996: 272), which contributed to inciting genocide and sparking the Second World War.

For all their emphasis on propaganda, however, the Nazis missed a possibly war-winning propaganda opportunity when they invaded the Soviet Union. Many of the ethnic groups in the western Soviet Union greeted the Germans as liberators, but the Germans' brutal racial policies quickly turned the people against them. If the Germans had used propaganda to persuade the liberated peoples to fight with them against Stalin's brutal government, the Germans might have won the war. In this case, however, racial beliefs trumped a golden propaganda opportunity (Blackstock 1958: 263–72).

Germany also failed when it directed its propaganda at the United States before the American entry into the war. The Germans sought US reporters who supported a neutral America to promote an anti-war policy. In June 1940, a Republican Congressman even asked the Nazis for $30,000 to help finance a series of newspaper advertisements headlined, "Keep America Out of War!" (Breuer 1988: 29–30). Propaganda is rarely a one-sided game and, even as the Germans tried to persuade the Americans to stay out of the war, the British were conducting a massive, ultimately successful campaign to persuade the United States to enter the war on their side. Many of Churchill's stirring speeches were, at least in part, aimed at a US audience to convince Americans that Britain was not doomed and was worth supporting in the bleak fall of 1940. Other factors from culture and ideology to trade and a shared history were important but propaganda played a significant role in demonizing Nazi Germany and making Britain appear to be the last, valiant outpost of Western Civilization in the minds of the American people.

As in the First World War, the British also used propaganda extensively against the Germans in the Second World War. By 1939, radio was a powerful instrument for propaganda and the British relied on it heavily. One of the finest practitioners of radio propaganda was Sefton Delmer, an Australian whose father had lectured at the University of Berlin. Fluent in German, Delmer broadcast on the BBC's German service. He exhibited an optimistic attitude to show the Germans that the British were confident in the summer and fall of 1940, as England stood alone against the Nazis. For example, in mid-July he broadcast an offer to teach German soldiers some helpful English phrases for their upcoming invasion of Britain, including "Channel crossing," "the boat is sinking," "the water is very cold" and the conjugation of the verb, to burn. "And now I suggest that you learn another phrase: *Der SS Sturmfuehrer brennt auch ganz schoen*... the SS captain is al-so burn-ing quite nice-ly" (Breuer 1988: 35–6). Delmer's choice of topic was far from random. Based on the principle that propaganda should be reinforced by other means, his story was supported by rumors planted through double-agents and the media that the British had developed a "devious method to set the Channel afire" when the Germans invaded (Breuer 1988: 35–6).

Delmer soon realized, however, that the BBC's German service sounded like one anti-Nazi German refugee talking to another anti-Nazi German refugee, and would have little effect on most Germans. The key, he realized, was to reach the millions of Germans who supported Hitler and the war, not the few who already opposed the dictator. Delmer's suggestion to use deception to reach this majority of loyal Germans, however, did not sit well with the BBC's executives. The BBC was the voice of truth, trusted around the world as an independent source of news. It was not supposed to be a mouthpiece of the British government. With the BBC's credibility at stake, Delmer gave up on his plan.

Then, in mid-1941, Leonard Ingrams sent for Delmer. Ingrams was an official in the Ministry of Economic Warfare, which coordinated anti-Axis subversion. He proposed starting a right-wing station in response to the recent German establishment of a left-wing station, *The Workers Challenge*, in which Germans speaking perfect English tried to undermine the British war effort. "Little old ladies in Eastbourne and Torquay are listening to it avidly, because it is using the foulest language ever," Ingrams said. "They enjoy counting the F's and B's." The British decided to reply in kind.

The gifted Delmer became the chief of this "black" radio station. Subtlety was crucial to success. Instead of broadcasting information from London that Germans would largely ignore because it would be seen as propaganda and, therefore, would lack credibility, Delmer gave the impression that the station was operated by patriotic Germans somewhere in Nazi-occupied Europe. The Gestapo, however, appeared to be after these loyal Germans. More Germans would listen if it appeared that they were listening to information from a German, not a British, source, especially if they thought that they were not supposed to hear it. The chief character in the charade was a "crusty, foul-mouthed, old Prussian" called *Der Chef* (The Chief). *Der Chef*, however, was neither old nor Prussian. He was 34-year-old Paul Sanders, a Berliner who fled Germany in 1938 and then joined the British Army when the war started. *Der Chef* was contemptuous of Churchill, but also of the high-level officials around Hitler, although he was fiercely loyal to *der Feurher*.

The station began broadcasting on 23 May 1941, as *Gustav Siegfried Eins* (George Sugar One) a signaler's phonetic wording for GS-1, which was a random identity meant to add to the station's mystique. After Rudolf Hess flew to Scotland, *Der Chef* broadcast:

> First, let's get this straight. This fellow [Hess] is by no means the worst of the lot. He was a good comrade of ours in the old days. But like the rest of this bleeping clique of cranks, megalomaniacs, string pullers, and parlor Bolsheviks who call themselves our [German] leaders, he simply

> has no guts for a crisis. As soon as Hess learns a little of the darker side of the developments that lie ahead, he loses his head completely, gets a white flag, and flies off to throw himself and us all on the mercy of that bleeping flat-footed bastard of a drunken old Jew Churchill.
>
> (Breuer 1988: 93–4, 101–3)

In closing, *Der Chef* cautioned the secret cells of a non-existent underground army that the station would be silent for part of each day so they could move to a new locale, since, they claimed, the Gestapo was chasing them.

Der Chef once reported on the court martial of a German officer whose blunders had led to the loss of a battalion; "If this goes on, we'll soon have all our officers before court martials... To err is human, isn't it? Most of our officers err once in a while and they cannot help it if it results in the annihilation of a battalion. Such is war!" (Farago 1958b: 671). Later, as Hitler's legions faltered in the outskirts of Moscow, *Der Chef* ranted against those who had failed the Fuehrer. In writing scripts, Delmer carefully avoided naming Hermann Goering, Heinrich Himmler and other top Nazi leaders, since that would have sounded like enemy propaganda. The station's targets were the lesser-known Party leaders, the "spineless bastards" conspiring behind Hitler's back. British intelligence gleaned the names of lower-level officials from POW mail and Reich newspapers. *Der Chef*, for example, criticized those who made big profits on shipments of winter clothes, which delayed their delivery to frontline troops, and was furious that blood was taken from Russian prisoners for transfusion to wounded German soldiers without being tested for venereal disease (VD). Therefore, *Der Chef* ranted, "our wounded boys" had VD and when they returned home they would "pass it along to their wives." This, Delmer hoped, would lead to rather cold homecomings for many German soldiers, thus lowering morale (Breuer 1988: 111–12).

Eventually, however, the story that the Gestapo was after the station began to wear thin. The decision was made to have the Gestapo kill the staff on the air. The scene was set, gunfire erupted and the station's last broadcast was sent into the ether. The radio operators, however, were bored and broadcast the incident twice by accident, thus making it appear that the Gestapo had killed *Der Chef* twice. Luckily, the Germans were accommodating. They were jamming the first signal, so few heard the broadcast twice (Becker 1958: 675). Even with such gaffes, the operation was extremely credible. The Americans, for example, did not even know that the station was a British operation until June 1942 when two US intelligence specialists finally figured it out (Farago 1958b: 672).

Another effective Allied propagandist was Lieutenant Commander Ralph G. Albrecht, who broadcast 309 times to Germany between 8 January 1943 and the end of the war. His special emphasis was the navy. "I have just heard that 16 new admirals of the German Navy have been appointed," he broadcast. "That makes a total now of two grand admirals, five general admirals, more than 150 admirals! ... Just what do your admirals do? ... Are they aboard warships? Well, hardly – because there aren't enough warships for that." Since "there are only

30 German warships of more than 1,000 registered tons… five admirals would have to divide one flagship between them. You certainly could say in this case that you can't see the fleet for the admirals.… In April [1942] there were actually more admirals launched than U-boats" (Daugherty 1958d: 496). The broadcasts must have hit a Nazi nerve because in September 1943 the Germans stopped publishing promotion lists in the newspapers, which was where the Allies got the information for the satiric broadcasts.

Propaganda, however, can backfire, emphasizing the need for astute analysis of a situation. In February 1942, the Germans introduced on the air a "plump, middle-aged American-born (Atlanta) woman named Jane Anderson." The British soon christened her Lady Haw-Haw, a female counterpart to Lord Haw-Haw. Anderson was on the air four times a week and quickly became Goebbels's favorite. In March 1942, Anderson sought to set her US audience straight about supposed German food shortages by reporting on her visit to a Berlin cocktail bar:

> On silver platters were sweets and cookies. I ate Turkish cookies, a delicacy I am very fond of. My friend ordered great goblets full of champagne, into which he put shots of cognac to make it more lively. Sweets, cookies, and champagne! Not bad!
>
> (Breuer 1988: 140–1)

Upon hearing the broadcast, British propaganda experts were delighted. The next night, the British radioed her report, translated into German, back into the Reich. No editing was required. The German people had begun to feel the pinch of food shortages and the effect of the turnaround broadcast was immediate. Anderson's radio career was over; such can be the power of a broadcast reaching an unintended audience.

Propaganda was also crucial to the underground movements in Nazi-occupied Europe. By the spring of 1942, for example, there were dozens of clandestine newspapers in France, such as *Le Franc-Tireur* and *Combat*, with a total circulation of about 200,000 a week. Much of the news in the newspapers was from the BBC, to which many people had already listened, and the cartoons were cut from old magazines, but the editorials were powerful and helped recruit new members, raise money and gather information. The newspapers could also be effective on specific issues, such as when they helped block the German attempt to recruit French workers to work voluntarily in the Reich. The newspapers were printed wherever there were presses. One underground paper was printed in a hotel's basement. The hotel was entirely occupied by German officers, who paid no attention to the rumble of the press in the basement, since the press also printed the hotel's menus. The newspaper operation was eventually moved for fear the Germans would notice that the newspaper's typeface was identical to that on the menus from which they ordered their meals every day (Breuer 1988: 141–2).

As with the Nazis, Soviets and British, the Chinese Communists used extensive propaganda in their wars, first against the Japanese, then in their civil war, and finally during their invasion of Tibet. At first, Chinese propaganda against the Japanese was crude. For example, exhortations to "Overthrow the Emperor" were ineffective. The Chinese, however, then started using Japanese POWs as propagandists, forming an Anti-Enemy Section that enforced good treatment of prisoners. In fact, taking a page from Benno Frank's *modus operandi*, if the prisoner wished, the Chinese would release the Japanese prisoner back to their own lines, often with gifts and after a farewell party. In 1940, the Chinese also established the Japanese Workers' and Peasants' School in Yenan to train POWs in psyops. Chinese psyops were extremely effective. Of one group of 3,000 Japanese prisoners, for example, 300 agreed to be trained for psyops against their former comrades (Mohler and Livingston 1982: 367–8).

With the defeat of Japan, the Chinese Communists turned their propaganda on the Nationalists. The Communists often entered cities with posters and paint pots, as well as their weapons, with orders to "Paste-shoot, shoot-paste" (Mohler and Livingston 1982: 369). Communist propaganda contributed to large-scale Nationalist defections and helped win the war. After the civil war, the Chinese invaded Tibet in 1950 and, once again, psyops were crucial. On 18 October 1950, for example, the Chinese approached the Tibetan fortress of Chambo. The 3,000 defending Tibetans expected a conventional attack but instead, as night fell, they heard explosions as star shells and rockets lit the sky. Rumors spread of an immense impending Chinese attack. The garrison panicked and the commander fled. Not a shot was fired and the fortress fell to the invaders. Chinese propaganda, which had been so effective against the Japanese, Nationalists and Tibetans, however, failed when the Chinese faced the Americans and UN forces in Korea. The Chinese lacked insight into the Western mind and much of their propaganda seemed crude to Westerners (Mohler and Livingston 1982: 366–73).

After the Second World War, anti-colonial liberation movements often used psyops and propaganda. In fact, almost no liberation movements actually defeated the Western colonizers on the battlefield. They won by using psyops. In Algeria, Angola and Namibia, the liberation movements were actually militarily defeated even as they achieved independence. The battle to capture the language and images of a conflict were crucial to victory. Many anti-colonial revolutionaries attempted to link their revolution to George Washington, Thomas Jefferson and the American Revolution, even as they sought to create negative images of their opponents. Even something as simple as persuading the media to call rebel executions, "assassinations" was a crucial step in transforming a rebel group into a recognized political entity on a moral level with a state. Conversely, any attempt at government control was called "repression." Charges of repression were reinforced by guerrilla-founded front organizations in the West, especially human rights groups, which criticized any and all government attempts to combat rebels (Radu 1990: 118–43).

Propaganda and psyops, however, also can be used against rebels. Philippine President Ramon Magsaysay waged a successful campaign in the early 1950s against the Huks. He coupled propaganda with other measures, such as punishing troops who abused the public, embedding civil affairs officers with all army units and, for the cost of about five cents, offering anyone the opportunity to telegram his office to report Army brutality, which would be investigated within 24 hours. In large part by making his government appear credible, trustworthy and reliable, Magsaysay's government defeated the Huks (McLaurin *et al.* 1976b: 400–2).

Psyops were also used in conventional conflicts after the Second World War. During the 1956 Suez Crisis, as Egyptian President Gamel Abdul Nasser broadcast speeches exhorting colonial peoples to rebel against the Imperialists, the British and French attacked Nasser with forged pamphlets stating that Egypt sought to control all of the Middle East's oil and spread stories that ex-Nazis were operating Egyptian concentration camps in an attempt to weaken international support for Nasser (Dorril 2000: 624–5). The Arab News Service was actually run by the British Foreign Office's IRD. The Service was a perfect vehicle for propaganda, since it was taken by almost every newspaper in the region, sometimes free of charge to encourage its use. The British also controlled radio stations throughout the Middle East that broadcast anti-Nasser propaganda. In the late 1940s and early 1950s, MI6 also owned Sharq al-Adna, the Arabic name for the Near Eastern Arab Broadcasting Station. Sharq transmitted from Cyprus and was the most popular station in the region, based on strong music and drama programs. The British relied on a staff of 150 Arabs, mostly Palestinians, who were gifted at getting into the minds of their Egyptian brethren (Aldrich 2002: 482–5).

Psyops can sometimes carry a high cost, however. Everyone knew that the Arab News Service building in Cairo was a front for British intelligence operations. In August 1956, the Egyptian secret police raided the building and arrested several British operatives and 12 Egyptians. Mustapha el Hebawi, head of Egyptian State Security, gleefully rubbing salt in the wound, told the British Ambassador that three Israeli spies had already been dealt with: two executed and one had committed suicide. London feared Egypt would execute their MI6 propagandists, but Nasser showed restraint and only executed the Egyptians entangled in the British operation (Aldrich 2002: 483).

The British were not the only ones engaging in psyops in the 1950s. As the French were leaving Hanoi, the Viet Minh had secret plans to welcome their troops into the city. The CIA suggested printing a fake Communist manifesto ordering everyone in the city except essential hospital employees to be out on the streets not for a few hours, but for a celebration lasting an entire week. Transportation, electric power and communications would cease in what would, in effect, be a seven-day general strike. The manifesto looked so authentic that Communist leaders in the city worked hard to make sure turnout was 100 percent. A last-minute radio message by the Communists outside the city to ignore the manifesto was taken as a French attempt at counter-propaganda. It took the Viet Minh three days to restore public service (Lansdale 1976a: 762–3).

Later, the Americans published the 1955 predictions of famous soothsayers in Vietnam. The books were sold for a nominal price, since if they were free people would have suspected propaganda. The soothsayers predicted troubled times for those in Communist areas, infighting among the Communist leaders and various other calamities. The book was a success and even became a bestseller in Haiphong. Some of the predictions, such as the bloody suppression of farmers opposed to land reforms and splits in the Politburo, even came true (Lansdale 1976a: 762–3).

North Vietnam and her allies also routinely used propaganda. An East German film about captured US airmen in North Vietnam was especially effective. The film was shot in a documentary format, so that it appeared objective and credible, and not just propaganda. Images were also used effectively. Scenes of bomb damage and injured women and children were intercut with film of smiling US pilots returning from missions. An interviewer then asked the pilots whether the United States had declared war on North Vietnam. Since there had been no declaration of war, the pilots said "no" and thus appeared to be war criminals, not POWs. The film also stressed that the airmen had their side-arms with them when they were captured, yet, contrary to the US Military Code of Conduct, had not used them. The implication was that the pilots were brave when they were dropping bombs on women and children, but were cowards when face-to-face with their victims. Following US tradition, the pilots also said they had no interest in politics. The pilots' statements only reinforced the idea, common overseas, that Americans are politically naïve. After each long interview with a pilot, the commentator then stated that all the information US military personnel are supposed to provide under their Code of Conduct is name, rank, serial number and date of birth. This statement suggested that the pilots were not even living up to their own code (Preston *et al.* 1976: 855–62).

During the Cold War, the East Germans were part of a massive Soviet psyops organization. The Soviets called such operations "active measures," which were conducted by the KGB and ranged from running agents of influence (influential people who supported the Soviets) to forgeries and the dissemination of bogus news stories. In 1981, for example, the KGB published books on Northern Ireland that included a forged letter linking British Prime Minister Margaret Thatcher and South Africa's apartheid regime (Andrew and Gordievsky 1990: 617, 628–32). White or overt propaganda was produced by many Soviet organizations, including the Ideology Department of the Communist Party of the Soviet Union, Radio Moscow, Pravda and TASS. In 1988, for example, an official Soviet booklet, *About the CIA in Asia*, alleged that the CIA had orchestrated the assassination of the US ambassador to Afghanistan, Adolf Dubs, to engineer a rift between the United States and Afghanistan. Dubs was actually kidnapped by terrorists in Kabul and the Afghan police opened fire when told to by their Soviet advisors, killing the ambassador even though the United States had wanted to continue negotiations. Another Soviet booklet, *Once Again About the CIA*, published in 1989, blamed the CIA for the 1978 Jonestown massacre, where more than 900 cult members committed suicide or were killed on orders of their insane leader, Jim Jones. When the United States protested the charge, the Soviets merely claimed that the booklet was part of the Soviet's new greater openness (Romerstein 1990: 53).

Soviet gray propaganda was conducted through foreign communist parties under Moscow's control and front organizations, such as the World Peace Council, which argued that Soviet nuclear weapons were no threat to peace, while focusing on disarming the West of its nuclear arms (Romerstein 1990: 36, 50). One of the more incredible Soviet-initiated stories that gained credence was a story about Americans importing Latin American babies to harvest their organs for transplants. The story was started by a Soviet front organization, the International Association of Democratic Lawyers, which was supported by the French Communist Party. The Association even managed to convince the European Parliament (EP) in September 1980 to initiate an investigation. The vote to investigate was taken while most EP members were absent (Romerstein 1990: 48).

The Israelis also used psyops. During the 1973 October War, for example, Radio Israel broadcast that Syria had executed a Druze lieutenant colonel for neglect of duty. Although a minority, Syria wanted to retain the loyalty of the Druze to avoid Syrian Druze siding with the Druze community in Israel. The Israeli operation must have been effective, because Syrian media devoted a great deal of time and space to debunking the charge. The lieutenant colonel, Syria claimed, had been killed fighting the Israelis. Thereafter, Syria threatened heavy punishment to anyone passing on Israeli rumors, and a Druze leader was invited to speak on Syrian radio and television. Even eight months later the Syrians were still concerned about the story. On 7 July 1974, the remains of Lieutenant Colonel Rafiq Halawah, the officer in question, were taken from a military hospital and buried in Damascus at the Cemetery of the Martyrs of the October War. The Syrian chief of staff, minister of defense and several military commanders attended the funeral, far more than probably would have been in attendance if Halawah had not been mentioned on Radio Israel (van Dam 1982: 358–62).

Propaganda continued to be a powerful tool in the 1990s. A young Kuwaiti woman testified before Congress in October 1990 about Iraqi atrocities just after the Iraqi invasion of Kuwait. Her description of Iraqis taking babies out of incubators in a hospital attracted widespread media coverage and helped paint the Iraqis as ruthless, brutal invaders. It was only later that it became known that the young woman was the daughter of the Kuwaiti ambassador to the United States. Her testimony had been created by a public relations firm, Hill and Knowlton, which had been paid by the Kuwaiti royal family (Severin and Tankard 2001: 115).

In 1992 Serbian President Slobodan Milosevic used propaganda in his bid for reelection against Prime Minister Milan Panic. Milosevic, "the butcher of the Balkans," had been elected president in 1989 by playing upon ethnic divisions between Serbs, Croats and Muslims. In the 1992 campaign, Serbs were encouraged to believe that if Milosevic lost, foreign powers would impose Muslim rule on the country, Serbian women would be forced into harems, and Serbs would be placed in concentration camps. Milosevic used state-run radio and television to deliver his message and to distort his opponent's image, labeling Panic a traitor and a CIA agent. An independent radio station, B-92, opposed Milosevic, but the government shut down the station. International pressure, however, led to the reopening of the station. Radio Free Europe and the VOA rebroadcast its signal, which was then piped onto the Internet, ending the state's monopoly on information. Even so, Milosevic won with 56 percent of the vote, although the voting was marred by irregularities (Adams 1998: 90).

China, as it did during its war with the Japanese and its civil war, still considers psyops and propaganda to be central to its global strategy. For example, it tried to stop the release of *Kundun*, a Disney motion picture about the Dalai Lama, exiled ruler of Tibet, which China conquered in 1951. Chinese pressure appeared to fail, in that *Kundun* was released, but the pressure meant that anyone wanting to film in the People's Republic shelved plans for any films with anti-Chinese themes. The same phenomenon has occurred in academia. Scholars who want to conduct research in China had better not pursue anti-Chinese themes or they will have great difficulty gaining access to Chinese sources. The Chinese also have pressured Western companies to produce films and books to distribute in US schools about their projects in China. By portraying Sino-American cooperation, the films and books always offer a positive view of China. The Chinese have also recruited major political figures ranging from Henry Kissinger, Cyrus Vance and Lawrence Eagleburger to Dick Cheney, George H.W. Bush and Brent Scowcroft to represent their interests in Washington or to work promoting companies with close ties to China (Adams 1998: 253–7). Following China's lead, some Arab states are now looking into ways to air programs or advertisements on US television stations to promote their agenda and improve their image (Murphy 2003: Online).

Although psyops and propaganda have only rarely used altered images in the past, they would be a powerful addition to some, if not all, such operations. Words are powerful, but people rarely question the validity of photographs, video and films, especially when the image reinforces a belief they already hold, and almost never evaluate the emotions aroused by images. Past examples of psyops and propaganda provide only a hint of the vast possibilities altered images and video offer for those who produce propaganda and practice the ancient art of psychological warfare.

5

DECEPTION IS A MANY AND VARIED THING

Leaders make decisions. Decisions are based on information. Therefore, if the information upon which an adversary makes decisions can be distorted, altered or even just garbled, it can give a significant advantage to the deceiver. Operations that seek to deceive adversaries in international relations and war have been extraordinarily effective throughout history. Modern image-altering techniques can be used to support such operations making them, if anything, even more effective.

Writing about the effectiveness of deception in politics, diplomacy and war, however, is inherently difficult. To begin with, researchers are only aware of a subset of deception operations that have been made public. Since information about such operations is usually provided by those involved, there is a tendency for only those operations that have succeeded to enter the public domain. Few wish to put their name on a book describing a deception in which they participated that failed. Furthermore, institutional gatekeepers also can conceal failures. Making deception even harder to research is the fact that there must be deception operations that succeeded, yet are still secret. For all of these reasons, as well as the inherently secretive nature of intelligence work, it is difficult to distinguish truth from fiction in researching the effectiveness of deception operations.

Even taking into account the bias toward reporting successes, the available evidence strongly suggests that all forms of deception are extremely, even extraordinarily effective. Barton Whaley, for example, in *Stratagem: Deception and Surprise in War* composed a 600-page appendix listing all of the known twentieth-century military deceptions. He found that almost all of them worked (Epstein 1989: 296 note 2). Furthermore, the goal of deception is usually surprise, which has long been recognized as fundamental to success in military operations. Surprise changes the ratio of casualties in favor of an attacker from 1:1 to 5:1, and doubles the combat power of those who achieve it (Gerwehr and Glenn 2000: 37–8). Frederick the Great wrote, "It is pardonable to be defeated, but never to be surprised" (Brown 1975: 415). In the twentieth century, deception became crucial to achieving surprise, even being called the "midwife of surprise." Whaley found that in 68 major battles between 1914 and 1967, surprise became more and more reliant upon deception (Gerwehr and Glenn 2000: 38). Without the use of deception, surprise, at least in conflicts since 1914, was only achieved about half the time (Mullen 2005: Online). There is no reason to believe that deception and its relationship to surprise will be any different in the twenty-first century.

It is a comforting myth that you cannot fool someone with the same trick twice, let alone thrice or even more times. In reality, deception can work against the same opponent repeatedly. All that is required is to offer a choice, such as either/or, left or right, up or down. The military call it the principle of alternative goals. General William Tecumseh Sherman's 1864 drive to Atlanta is a classic example of a serial deception. Sherman's logistics were tied to a single railway and he had to attack along that railway. The Confederates knew this fact, yet in every attack, save one at Kennesaw Mountain (where there was little room for an attack on either side of the railway and Sherman had to order a frontal assault on the Confederate positions), Sherman surprised the Southerners. Sherman had the left–right option, and he used it to deceive and repeatedly surprise his enemy by attacking to the left or right of the rail line (Bowyer 1980: 40–1).

Like surprise, which is often the goal of deception, the use of deception strongly correlates with casualty levels. Deception can significantly shape a battlefield so as to greatly increase the probability of victory and lower casualties on both sides. Whether deception is used significantly influences whether a war becomes one of attrition or of maneuver. The less deception is used, the more a conflict will be a battle of attrition; the more that deception is used, the more a conflict will be based on maneuver. The Allied landing at Normandy, the US island-hopping campaign in the Pacific during the Second World War and General Douglas MacArthur's landing at Inchon during the Korean War are examples where the use of deception led to a war of maneuver that avoided what could easily have turned into long, bloody wars of attrition. If there is little use of deception, such as on the Western Front during the First World War, during the Allied advance after D-Day or the hill battles in Korea in 1951–3, there is far more likely to be attrition warfare, with much higher casualties. For example, MacArthur's deceptive campaign of island hopping in the Pacific cost far fewer American lives than the deceptionless drive across Europe from Normandy to Germany.

One of the most famous deception operations was Bodyguard, devised to deceive the Germans about the timing, location and scale of the Anglo-American-Canadian landings at Normandy in June 1944. After achieving surprise, however, the large-scale use of strategic deception ended relatively soon after the landings. The British Special Plans Branch concluded:

> The overwhelming success of the Fortitude operations [part of Operation Bodyguard] was followed by a period of partial and/or indifferent success during the remainder of the campaign on the Continent. No over-all [*sic*] strategic deception was attempted. Tactical deception, despite a record of successful minor manipulations of enemy intelligence, was characterized by a succession of wasted opportunities.
>
> (Brown 1975: 805)

The end of strategic and tactical deception operations led to a significant increase in casualties and far more of a war of attrition and far less a war of maneuver. When the Allies used deception to deceive the Germans about the Normandy landings, the Americans suffered 133,326 casualties (34,133 dead) and the British and Canadians 83,825 casualties (16,138 dead). During the march across France to Germany, when little, if any deception was employed, the Western Allies suffered 418,791 US casualties (86,000 dead) and 107,000 British and Canadian casualties (25,000 dead). In the later, albeit longer, period, when deception was rarely used, US casualties were more than three times that of the hazardous landing period, when deception ruled and higher casualties were expected than actually occurred (Brown 1975: 796–7).

Even on the Western Front in the First World War, deception could have been used to ease the costs of attrition warfare and increase the chance for maneuver. For example, as Wood (1972) showed, in 1917 four Canadian divisions used creative and, in part, deceptive tactics to take Vimy Ridge after the French had suffered 150,000 casualties while attacking to take it with conventional tactics.

At the tactical level, deception in the form of camouflage, decoys and electronic measures are far cheaper, faster to employ and more effective at lowering casualties than attempting to make the real target tough enough to survive being hit. It is far better to try to avoid being hit, than to try to withstand being hit. In short, deception is "as valuable as armor, speed, or firepower" (Gerwehr and Glenn 2003: 59).

Once deception plans are played out and battle is joined, however, battles are won by will, force, blood, sweat and toil. During the landings on the beaches of Normandy, for example, the British, Canadian and American soldiers and sailors were the ones who had to fight their way through the German defenses. Absent deception at the strategic, operational and tactical levels, however, those defenses would have been far stronger. The great deception plan that shrouded the invasion shaped the battlefield to ensure that the German forces were well dispersed and caught by strategic, operational and tactical surprise. As Dudley Clarke, the top British deception expert during the war, wrote in his unfinished memoirs, "the secret war was waged to conserve rather than destroy; the stakes were the lives of the frontline troops, and the organization which fought it was able to count its gains from the number of casualties it could avert" (Latimer 2001: 311). Deception cannot guarantee victory, although it certainly can tilt the battlefield significantly in the direction of the deceiver.

The deceiver's task is made infinitely easier because it is extremely difficult to perceive reality accurately. Humans are deceived every day in many ways that we rarely, if ever notice. If looked at in a certain way, almost all of psychology deals with deception. Studies of perception analyze how we perceive the world, yet almost all such studies find that humans rarely perceive reality accurately with any of our five senses. Some physicists even question whether time and space actually exist in any form remotely like how we perceive them (Greene 2004: R4).

Errors in attribution, which can be seen as deceptions, are also extremely common as we seek to attribute meaning to the behavior of others and ourselves.

Regardless of reality, we tend to attribute success to internal factors, but failure to external factors (Baron *et al.* 1995: 59). We tend to believe, for example, that if we succeed on a test it is because we are smart or studied hard, yet if we fail, we attribute the failure to the teacher being incompetent or the test being unfair. The same is true of nations. Most Americans believe that the United States was victorious in the World Wars because of a vast industrial base, patriotic, hard-working citizens and brilliant leaders, all factors internal to the United States. Yet most Americans believe that the United States lost in Vietnam because it was fighting a truly nationalistic movement, the South Vietnamese government was corrupt or other factors external to the United States.

Our allegiances, whether to nations or to schools, can also lead to deception. The 1951 Princeton-Dartmouth football game, for example, was a grudge match with fistfights and many injuries. After the game, two psychologists showed films of the game to students from each school, asking them to act as scientists and note each infraction and who was responsible for each infraction. Princeton students were much more likely to see their own players as victims of foul play, while Dartmouth students were just as likely to see their own players as victims of illegal aggression (Myers 1996: 8–9). The same, of course, applies to nations, where most citizens invariably see international crises as caused by the other state, rarely by their own.

We are also deceived because we are an optimistic species who believe in a just world. Most people believe that only evil people are punished and that good people usually do well. Such beliefs often lead to self-deception, where a more objective view would have been far healthier. In 1943, psychologist Gordon W. Allport analyzed 200 life histories written by refugees from Nazi Germany. Not one of the refugees, at first, believed that Hitler could cause such catastrophes as the Holocaust and the Second World War. In 1932 they hoped and believed that Hitler would never come to power. In 1933 they hoped and therefore believed that he could not put his threats into effect, and in 1934 they believed that the nightmare would soon pass, when, in fact, it only worsened. Such optimism, however, is extremely resilient. The 1989 San Francisco earthquake shook residents' optimism that they were less likely than their peers to be injured in a natural disaster. Their optimism, however, returned in just three months (Myers 1996: 56).

The same optimism extends to leaders making decisions, for which, invariably, they believe they are correct. Almost every leader who has led a state into war has believed that their country would win and, usually, quickly (Blainey 1988: 35–56). Half were wrong about winning and almost all were wrong in believing that the war would be brief.

Optimists are also more likely to get top jobs, which can lead to deception and, sometimes, disaster. It is extremely rare for a pessimistic leader to rise to the top of any organization, whether a democracy, dictatorship, corporation or the military. During Japanese war games to test the 1942 Midway operation, for example, the senior Japanese umpire, Rear Admiral Matome Ugaki, overruled other umpires when their rulings threatened to disrupt the Japanese plan. In one game,

"US planes" broke through the Japanese fleet's defenses and "sank" the fleet aircraft carriers *Akagi* and *Kaga*. Lieutenant Commander Masatake Okumiya ruled the sinkings "good," but the optimistic Ugaki overruled him. In the actual battle in June 1942, the Americans sank not two, but four Japanese aircraft carriers (Prange 1982: 35–6).

The heuristics or mental shortcuts we use to solve problems also can lead to deception. Leaders, for example, often use historical analogies to analyze foreign policy crises, even though history never repeats exactly. Therefore, leaders are deceived into thinking a current crisis is a repeat of a previous one and use lessons from the past to shape policy, even though the situations are different. During the Gulf War, for example, as Macdonald (2002) argued, President George H.W. Bush relied on an analogy to Hitler to frame the crisis. Bush reacted as the analogy prescribed; he used prompt military force to eject Iraq from Kuwait. The analogy suggested that Saddam Hussein should have then backed down and avoided confrontation. Hussein did not, leading Bush confused; his analogical frame did not fit the new situation. He had been deceived in his use of an analogy.

Such errors related to heuristics are just one small part of a vast literature by Jervis (1976), Vertzberger (1990) and others on misperception and errors in foreign policy decision-making. Such errors can be seen as forms of deception in that the leaders were deceived, either by their own senses and minds or intentionally by an adversary. The obstacles leaders face in making effective decisions are based on common human errors and deceptions, but also, in part, are caused by the fact that international relations and war are nonlinear. A small cause can have a large effect; a large cause, a small effect. Feedback is delayed and often ambiguous. Problems and their possible solutions are ill-defined and often change. Because of this complexity, research on attempts by foreign policy makers to learn paints a bleak picture. One study found that foreign policy experts were no better than amateurs in predicting even general future events in international relations. Experts only did better than laymen when they focused on extrapolating long-term trends (Macdonald 2000: 17). In short, perceiving reality accurately in war and foreign policy is extremely difficult, even for experts. We all, however, make common errors in problem solving, whether in our jobs or social lives, just as leaders do when addressing questions of war and peace.

Even though deception is effective and common, it is often condemned. Sir Walter Scott warned in *Marmion*, "Oh, what a tangled web we weave, when first we practice to deceive!" His words echo a long history of opposition to the use of deception because it is viewed by some as unfair, ungentlemanly and even unseemly. Even one of history's most famous deceptions, the Trojan Horse, was criticized when the ruse was proposed. In 1183 BC, Neoptolemus, the "battle-eager" son of Achilles, opposed the ploy. He argued, "Brave men meet their foes face to face! Away with such thoughts of guile and stratagem!" (Bowyer 1980: 19).

Even a successful deception does not lead to guaranteed acclaim, even if it wins a war. By 1901, after two years of conflict, the Philippine Insurrection was dragging on inconclusively and had claimed the lives of 4,000 US soldiers.

On 8 February 1901, Colonel Frederick "Scrapping Fred" Funston received intercepted dispatches from the rebel leader, Aguinaldo, reassigning several guerilla units to his secret headquarters, located somewhere on the island of Luzon. Funston devised a deception operation to capture Aguinaldo. Funston disguised 85 loyal Filipino troops as guerillas. He and four other US officers played the role of prisoners. On 6 March, the party headed for the rebel encampment. Using captured rebel stationary and forging the signature of one of Aguinaldo's most trusted commanders, Funston even kept the rebel commander informed of the location of the approaching "reinforcements." Aguinaldo responded with instructions as to what route the party should take. On 24 March, Funston and his party reached the headquarters and arrested Aguinaldo. Funston's force suffered no casualties, while only two rebels were killed and three wounded. The war was over. Funston received the Congressional Medal of Honor, but not for his brilliant deception operation that captured Aguinaldo and ended the war. The medal was for an act of bravery in combat the previous year. Moralistic journalists, in fact, "pilloried" Funston for his deceitful action (Bowyer 1980: 42–3).

Often deception and its oft-sought goal, surprise, are so looked down upon that some give it up even when it is achieved. At Tanga in East Africa in 1914, Royal Navy Captain F.W. Caulfield notified the unaware Germans of an impending British landing. The British then attacked. The Germans, now forewarned, fought back and the attack failed disastrously. Caulfield was not court-martialed. In fact, "he was promoted for his gentlemanly behavior" (Bowyer 1980: 43).

As with deception in general, the powerful, and sometimes almost everyone, condemn new weapons if they rely on deception for their effectiveness. In 1862, for example, Confederate Brigadier-General Gabriel Rains invented the landmine. In early May, a Union cavalry regiment was pursuing Rains' infantry as they withdrew up a road from Yorktown toward Williamsburg when the Union cavalry found themselves in the world's first minefield. Casualties were light, especially since there were only four mines, but the cavalry detoured a long way around, allowing the Confederates to escape (Bowyer 1980: 40). Union General George McClellan called the new landmines an infernal device and even Confederate General James Longstreet wanted Rains court-martialed. Rains' commander, General John Magruder, however, saw Rains' potential and had him transferred to the munitions bureau. Just as with the landmine, the British, who ruled the waves, condemned the introduction and use of the submarine and sea mine during the First World War. Such weapons relied on deception, and the weaker naval powers were bound to use them just as surely as the powerful were bound to condemn them (Massie 2003: 122–45).

The submarine and mine examples highlight the tension between the extraordinary effectiveness of deception and the tendency to condemn its use. Two factors affect the balance between deception's effectiveness and whether it is condemned and, therefore, affect how often deception is used in foreign policy and warfare: the use of deception domestically; and the relative power of a state.

As Bowyer (1980) argued, if deception in the form of plots, coups and assassinations is commonly practiced in a state's domestic politics, then deception is often used by that state's military in war. In large part, this internal/external relationship is due to the fact that for much of history the political leaders of states and empires, such as Alexander the Great, Julius Caesar, Belisarius and Napoleon, who employed guile to rise to the top and retain power politically, also served as generals who fought wars externally. There is still such a link in modern times, when many leaders, such as Mussolini, Hitler, Stalin and many leaders of Third World revolutions used deception to attain power and continued to use deceptive tactics once they were in office in diplomacy and during interstate wars.

Sometimes overriding this link between the internal and external use of deception is the relationship between a state's power and deceit. The strong usually rely on force, even though they could lower their costs and casualties by using deception. Being strong, they prefer to keep a contest solely one of strength, so as always to prevail. The weak, however, often use deception since they will lose if they rely solely on force to confront the strong. Slaves, for example, often used deception in the form of cultivating an image of being slow and stupid so as to avoid having to work too hard (Bowyer 1980: 5). Sportswriter Damon Runyon was right; "The race is not always to the fastest, nor the fight to the strongest, but that's the way to bet." Deception, however, ensures that the bet is never a sure thing. It is only as the strong decline that they begin to consider deception as a means of shoring up their waning power, especially if deception becomes more common in domestic politics.

Rome, for example, used deception when it was small and weak, but with Octavian and the founding of the Roman Empire, brute force came to the fore and deception fell from use. Rome had powerful, effective legions and weak, disorganized enemies, so there appeared to be no need for deception (Bowyer 1980: 22–3). As Rome declined, however, deception became more common internally. Between 53 BC and AD 96, 11 of 12 emperors were the targets of assassination; six were murdered and two committed suicide before they could be killed (Bowyer 1980: 357–9). This use of deception internally was mirrored by many who argued for the use of deception to shore up Rome's faltering legions on the frontiers of the empire. In AD 90, Frontinus wrote *Strategemata*, a typology of deception in military affairs, while Polyaenus's *Strategemata* in the AD 160s outlined 900 ruses in an attempt to help co-emperors Marcus Aurelius and Verus in the war with Parthia (Bowyer 1980: 23). All the suggestions, however, were too little, too late. By then, Rome was too far gone and the barbarians were at the gate; not even deception could save Rome.

Deception and intrigue were routine aspects of Byzantine politics. Of the 107 emperors of the Eastern Roman Empire over its 1,058 year (395–1453) existence, 23 were assassinated (Bowyer 1980: 357–9). This internal use of deception was reflected in its use externally, in diplomacy and warfare. Belisarius, a great Byzantine general, stopped a Persian–Saracen invasion in AD 531 by using a series of bloodless maneuvers, and in 542 stopped another powerful Persian army without even fighting a battle by relying on deceptive stratagems (Bowyer 1980: 24–6).

With Byzantium's fall and the rise of Charlemagne, however, deception and guile came to be seen as evil. The rise of chivalry meant that deceit was little used in Europe for about 800 years. Knights were powerful and could, until the advent of the long bow, use brute force to win, so deception as a tool of war declined. Domestically, politics was hierarchical and conservative, with less use of guile and deception. In the 1500s, as the church, chivalry and the dominance of the mounted knight declined (as the Battles of Crecy [1346] and Agincourt [1415] foreshadowed), secularism rose and the use of deception along with it. In Italy, especially, assassination and deception became the order of the day in domestic politics, which was reflected externally in warfare. This tendency to use deception was reinforced by the weakness of the Italian states compared with France. When France invaded Italy in 1494, Niccolo Machiavelli, voicing the age-old argument of the weak, stressed guile in *The Prince*, *The Art of War* and *Discourses* as a means to defeat the powerful invaders (Bowyer 1980: 34–5). He wrote in *The Art of War*:

> Though fraud in other activities may be detestable, in the management of war it is laudable and glorious, and he who overcomes the enemy by fraud is as much to be praised as he who does so by force... The prince should never attempt to win by force what he might otherwise win by fraud.
>
> (Epstein 1989: 229)

The 1500s through the 1700s were a time of religious war, domestic intrigue and political change that led to the creation of the modern state system. The strong were not powerful enough to rely solely on force, so deception became increasingly important in European warfare. In 1651, Thomas Hobbes wrote in *Leviathan*, "Force and fraud are in war the two cardinal virtues" (Bowyer 1980: 36). By the 1790s, Napoleon, who had learned deceit and deception in order to survive and then flourish politically during the French Revolution, often used guile on the battlefield as France conquered most of Europe. He censored the press before an invasion and then planted articles in the press to deceive adversaries about the timing, location and strength of his attacks. Once his campaigns were underway, he constantly changed unit designations, shuffled deployments and carried out feint attacks to deceive his enemies (Bowyer 1980: 38). By 1804, however, with France dominating Europe and Napoleon apparently secure as her leader, he abandoned guile for the brute force of France's numerous, experienced battalions, leading, at least in part, to longer casualty lists and his eventual defeat.

After the fall of Napoleon, the Europeans relied on brute force to gain and then rule their overseas empires. Deception was not required when the Europeans used repeating rifles, railways and steamships against spears, trails and canoes.

Domestically, democracy was spreading in Europe and assassinations, coups and plots were also in decline. Some, however, lamented the focus on force over deception. In 1869, British Colonel Garnet J. Wolseley wrote in *The Soldier's PocketBook for Field Service*:

> As a nation we are bred up to feel it a disgrace even to succeed by falsehood; the word spy conveys something as repulsive as slave; we will keep hammering along with the conviction that "honesty is the best policy," and that truth always wins in the long run. These pretty little sentences do well for a child's copy-book, but the man who acts upon them in war had better sheathe his sword for ever.
>
> (Bowyer 1980: 39)

With the First World War, brute force was still at its peak, with deception little used. Domestically, in most of the European states, whether democratic or monarchical, deception was also uncommon, leading to its lack of use in warfare. For the most part, therefore, the First World War offers few cases of deception in what quickly became a war of attrition. Even new technology, a traditional means of deceiving an enemy and achieving surprise, had little effect due to the way new weapons were introduced. The British introduced the tank on a small scale, while the Germans only gained a minor advantage from the first use of poison gas. The one great attempt to use strategic deception, the Australia–New Zealand Army Corps (ANZAC) landing at Gallipoli, failed due to inept commanders and the defensive strength of barbed wire, machine guns and artillery. Tactically, camouflage, a form of deception, did come to the fore, although mainly as an attempt to survive the hellish conditions at the front. Most uniforms had been brightly colored until then, but dun-colored uniforms, net-covered guns and mottled vehicles became standard in all of the armies involved.

One of the few successful examples of deception in the First World War occurred in the Middle East, where armies were often far less evenly matched, so the weak turned to deception to survive and, sometimes, win. In 1917, after two failed British attempts to take Gaza, General Edmund Allenby took command of British forces in the area. T.E. Lawrence, later to gain fame as Lawrence of Arabia, said Allenby turned deception from "witty hors d'oeuvres before a battle into the main point of strategy" (Bowyer 1980: 81). The Germans and Turks expected another straightforward attack seaward along the coast toward Gaza. Allenby chose to attack from inland toward Beersheba. The British, however, forged documents to convince the Germans that the attack would be along the coast. A British officer fleeing from a Turkish patrol dropped a haversack containing plans for the attack. The pack also contained money and valuables to lend credence to the story. After its loss, the British also issued orders for a search to find the pack to convince the Turks that the loss was, indeed, accidental and not a plant. Knowing the Germans would hear of British movements toward Beersheba, the British also leaked clues that there would be a diversionary attack on that city. The Germans, therefore, concluded that there would be an attack on Beersheba on 3 January 1918, but that it would be a small one. In reality, the main attack was against Beersheba on 31 December. As a final ruse, the British had the Royal Flying Corps drop cigarettes laced with opium over Turkish positions to drug the Turkish frontline troops. The Turks were "literally caught napping." The British took Gaza and, soon after, Jerusalem. Allenby used a similar feint-with-deception a year later on 19 September 1918 at Megiddo, and seven days later all of Palestine had fallen to the British (Bowyer 1980: 81–2).

Another area where deception was practiced to great effect during the First World War was in German East Africa. As Hoyt (1981) showed, Colonel Paul von Lettow-Vorbeck, a master of guerilla warfare, used a small force that never exceeded 14,000 men (3,000 German and 11,000 native troops) to deceive, outmaneuver and defeat Entente forces that, at times, outnumbered his force 10 to 1. He even defeated future South African Prime Minister Jan Smuts and a force of 45,000 men. Promoted to General, von Lettow-Vorbeck was never defeated in the field, only surrendering after the war was over.

For all of its use in the Middle East and East Africa during the First World War, it was only in the Second World War that deception came once again to the fore. Weak, outnumbered and alone, the British, especially, turned to deception to hold back the tide of Nazi aggression until the United States entered the war and marshaled its strength. The British had learned to use deception to retain their empire as their power waned and, by 1940, were well versed in its use. Realizing the importance of deception, Churchill said at the 1943 Tehran Conference, "In wartime, truth is so precious that she should always be attended by a bodyguard of lies" (Brown 1975: 389). When they entered the war, the Americans were too militarily weak to rely on brute force so they accepted, if somewhat reluctantly, the British lead in the area of deception. After D-Day, the Western Allies and, especially the Americans, had built up their forces and felt stronger militarily compared with the Germans. Therefore, just as Rome and Napoleon had done before them, the Allies relied on the biggest battalions to win, instead of on guile, and their casualties increased significantly (Latimer 2001: 3–4).

When force is the preferred method of making war and deception is rare in domestic politics, a preference for brute force is also shown in other areas. For example, in sports today in the United States there is a strong preference for power over finesse, let alone deception. Tennis stars power ace after ace past their opponents. The volley is dying. In golf, the long hitters dominate, rendering some of golf's most hallowed courses unusable for professional tournaments. Baseball has also seen the rise of the homerun, as scores approach that of high school basketball games. Sluggers are king. As a reflection of this preference for power over deception domestically in politics and sports, the United States has, since the Second World War, focused on firepower and brute force over deception and guile on the battlefield. Korea did witness the deceptive invasion at Inchon, but quickly bogged down into a war of attrition for barren hills. Vietnam, from start to finish, was a war of attrition, although the Hail Mary Left Hook in the Gulf War was a welcome change from the use of brute force. The Iraqi War in 2003, however, was a return to the use of brute force, although with some feints. Even so, the core strategy was to overpower the Iraqis with air power and armored columns culminating in "Thunder Runs" through downtown Baghdad. Deception is a way of thinking that can permeate an entire society. When brute force is preferred, deception is rarely used in any area of endeavor from sports to politics, diplomacy to war.

It would be a major failing to leave the impression that deception is always effective and has no drawbacks. Deception operations can require a long time to develop. For example, some of the phantom or notional units the British used to deceive the Germans about the Normandy landings had been created four years before in the North African theater. Deception operations also require a range of personnel with various specialties, and thorough coordination with military, political and economic policy. Without coordination, an attack can be a disaster and an amphibious landing, one of the most difficult forms of attack, a catastrophe. One such landing that lacked a supporting deception plan occurred on 19 August 1942, when a predominantly Canadian force landed at Dieppe, France. The goal of the raid was to help convince the Soviets that the Western allies were actually preparing for a cross-Channel invasion. For those with a more conspiratorial [or high-Mach (see Chapter 6)] mind, the British may have been trying to prove to the Soviets and Americans, who were pressing for an early landing in France, that a full-scale invasion in 1943 would end in disaster. For reasons that are still murky, there was no deception plan for the raid (Brown 1975: 81) and by 1942 the British certainly had the capability to undertake extremely effective deception operations as they had demonstrated in the North African theater.

On top of the lack of a deception plan to mask the Dieppe raid, there was a deception plan to convince the Germans that there would be a landing in France in 1942. The plan was designed to persuade the Germans not to transfer troops to the Russian front or southern Europe. The plan was successful, leading Hitler to issue Directive 40 in March 1942 ordering the construction of the Atlantic Wall (Brown 1975: 74).

On top of the lack of a deception plan for the raid and a deception plan suggesting that there would be an invasion, a German deception plan, Porto II, misled the Allies about the strength of the German forces in the Dieppe area. British intelligence, MI6, accepted that only 1,400 men of the 110th Infantry Division were defending Dieppe as they rested and refitted after brutal service on the Russian front. In truth, there were 5,000 infantry and 250 artillerymen defending the port with a reserve of three infantry battalions, an artillery battalion, a tank company, and an entire panzer division only six hours' march away (Brown 1975: 85).

It was the worst of all possible worlds: no supporting deception plan; lack of coordination between deception plans and actual military operations; and an effective enemy deception plan. The result for the Canadian landing force was far worse than decimation; the word "decimation" comes from the Roman practice of executing every tenth man when a legion disgraced itself. The Canadian assault force of 3,623 men suffered 60 percent killed, wounded or missing. Of the 4,963 Canadian troops involved, "the finest men of the finest regiments of Canada" (Brown 1975: 88), 68 percent became casualties. Most were killed before they got beyond the sea wall, and many before they even disembarked from landing craft. For two weeks, Canadian bodies washed up along 20 miles of the French coast around Dieppe. The Canadians were not the only ones to suffer. The RAF lost 106 aircraft and the Royal Navy 551 men, one warship and 33 landing craft. By comparison, the Germans suffered 391 casualties and lost 46 planes. At 5:38 pm on the day of the raid, Field Marshal Gerd von Rundstedt wired Hitler from his headquarters outside Paris, "No armed Englishman remains on the continent." He could have added that no armed Canadians were left on the continent either. Because of the lack of a deception plan, Colonel Oliver "The Fox" Stanley abruptly resigned as Controller of Deception, as the chief of British global deception operations was called. He is said to have "objected vehemently to the Dieppe operation and... refused to be associated with a regime that would risk butchering Empire boys for the sake of keeping Russia in the war" (Breuer 1988: 185; Brown 1975: 75–91).

A year later, Operation Starkey provided another example of a deception operation that was poorly coordinated with military operations. Starkey was designed to convince the Germans that the Allies planned to invade the Pas de Calais in 1943. The goal was to convince the Germans not to transfer forces to the Mediterranean or the Eastern Front, even though British forces in Britain were weaker in 1943 than they had been during the disastrous Dieppe raid in 1942 (Brown 1975: 318). Rumors were planted in the press that an invasion was imminent. The United Press reported that an invasion of Italy or France was likely in the next month, and headlines in the 19 August 1943 *New York Times* proclaimed, "ARMIES READY TO GO, SAYS EISENHOWER," and "ALLIES BID PEOPLES OF EUROPE PREPARE" (Brown 1975: 322–3). The deception planners even managed to convince the higher authorities to allow them to use the trusted and revered BBC to create conditions for what would appear to be an imminent French national uprising in support of the phantom invasion. The BBC broadcast "A" or alert messages to the Resistance. Such messages meant that the partisans should prepare to conduct certain operations. Normally, the "B" or action message would then be sent to trigger the actions, but under Starkey, the "B" messages would not be sent. Therefore, the Germans would hear many messages for the Underground and assume an attack was imminent, yet the Resistance would not attack; but only if the Resistance obeyed London (Brown 1975: 320–1). It was a risky undertaking. If the Resistance rose, it might be crushed, rendering it useless when the real invasion came.

The British soon realized that the political warfare campaign was dangerously out of sync with broader strategic goals. The Political Warfare Executive (PWE) sent a secret memo to the media, "Increased press speculation on the probability of the invasion of the Continent from the United Kingdom this summer may seriously prejudice our military intentions and may, at the same time, dangerously and prematurely raise the hopes of our own people and the people in the Occupied Territories." British press speculation about a possible invasion ceased, but continued in the United States, where the British had little influence on the press. The question became how to defuse the invasion rumors without telling the Germans that the rumors were a hoax. Only a devious mind could devise the simple answer. General Dallas Brooks, a Royal Marine at PWE, devised such an answer; blame the Germans. The British broadcast:

> Be careful of German provocations. We have learnt that the Germans are circulating inspired rumors that we are concentrating armies on our coasts with intentions of invading the continent. Take no notice, as these provocations are intended to create among you a situation where you may be caught. Lay low! Do only what you are told to do by the BBC.
> (Brown 1975: 322–4)

Apart from demonstrating the potential danger of deception operations running counter to military and strategic policy, the Starkey operation also showed that, while normally extremely effective, not all deception operations deceive the intended audience. The Germans realized Starkey was a fake and did not react to the threat of invasion Starkey was designed to raise. Rundstedt wrote, "The movements (the British) made, were rather too obvious – it was evident (they) were bluffing" (Brown 1975: 326). In fact, between April and December 1943, Hitler moved 27 of the 36 Wehrmacht divisions in the West to Russia, Sicily, Italy and the Balkans (Brown 1975: 326). Ironically, German forces in France were so reduced, the Allies probably could have invaded successfully in 1943.

6

HOW TO DECEIVE

Principles

A branch of applied psychology, deception "is equal parts art and science" (Gerwehr and Glenn 2000: 15). The principles of deception are the science; the application of those principles as stratagems are the art. Regardless of the deception operation, the principles are the same, while the stratagems vary endlessly, only limited by the creativity of the most devious and creative minds.

The word deception is derived from the verb to deceive, which the *Webster's II New College Dictionary* (1999) defines as "to cause to believe what is not true." This definition, however, is misleading in relation to deception operations. The goal of the deceiver in politics, diplomacy or war is not just to deceive an adversary, since perceiving reality accurately is so difficult and rare, as Karl von Clausewitz suggested: "A great part of the information obtained in war is contradictory, a still greater part is false, and by far the greatest part is somewhat doubtful" (Brown 1975: 462). The goal of the deceiver is to make an adversary perceive reality in a way that will help the deceiver by making the deceived *do* something the deceiver desires. As General Archibald Wavell, British Commander-in-Chief, Middle East, during the Second World War, wrote, the object is "to force the enemy to DO SOMETHING that will assist our operations, for example to move his reserves to the wrong place or to refrain from moving to the right place" (Capitalization in original; Brown 1975: 46). Devising a deception operation involves defining what the deceiver wants the adversary to *do* (not think), a target (the enemy decision-maker), a story (what the target is to be convinced is true), a plan (how to convey the story to the target) and events (the parts of the plan that will convey the story to the target). A timetable and scenario are developed to allow the enemy to know bits and pieces of information, which the British during the Second World War called "foodstuff," that is mostly true but points to an invalid conclusion. The deceiver's goal is to have the target act on the basis of that erroneous conclusion.

Contrary to what most people believe, deception is usually based on information that is true, yet is interpreted in such a way as to reach an erroneous conclusion. The deceiver, however, wants the target to reach that invalid conclusion. James Jesus Angleton, a CIA counterintelligence officer during the Cold War, said, "The essence of disinformation is provocation, not lying." The key is to lure the enemy into a trap with the truth. Everyone uses deception and disinformation based on valid information in their own lives, although they usually do not realize it. For example, many people leave lights on in their house when they are out to convince burglars to burgle elsewhere. The information, that the lights are on, is true. The conclusion the burglar draws is incorrect: that people are home. As long as your house is not burglarized, you have conducted an effective deception operation (Epstein 1989: 29). Angleton said, "deception must mimic reality enough to be accepted." Similarly, Wavell stressed that any deception must appear "plausible; it is unlikely to achieve its object unless the apparent intention disclosed to the enemy is sufficiently reasonable to have been included in the enemy's appreciation as one of the courses open to the British commander" (Brown 1975: 47).

There must be no simple way to check the validity of the facts or, much more importantly, the validity of the conclusion. Furthermore, the less time the target has to check a story, the easier it is to deceive them. A crucial element that makes it easier to implement and harder to detect deception is that information, in and of itself, can rarely provide proof of deception. For example, if a spy takes information from an enemy and the enemy is unaware of the theft, the stolen information is intelligence. The same information, however, if taken or given with the enemy aware of its loss or transmission, is disinformation. Therefore, the truthfulness or validity of the information itself is irrelevant in determining whether a source is reliable. The key, unfortunately, given its extreme difficulty, is knowing the enemy's state of mind and what he knows (Epstein 1989: 92–3). The problem is so difficult that Angleton suggested having two intelligence services: one to view each defector as a mole, the other to view each defector as a true defector. This duality is necessary because of the extreme difficulty in determining the authenticity of defectors, which cannot be done on the basis of the information they provide. A real defector may unknowingly provide false information fed to them by an adversary. Conversely, a mole may provide information that is true for years, if not decades, before providing one crucial piece of misinformation (Epstein 1989: 103–4). Agent provocateurs, for example, often provide valid information in hopes of gaining a significant advantage later. Such agents were used by the British and French in their colonies, Communist states internally and the United States against domestic Communist groups. Agent provocateurs sometimes even staged jail breaks or aided in attacks on their own forces, in effect sacrificing pawns for the larger goal of gathering information about opposition groups or in order to destroy the trust within the group. Once an agent provocateur has been exposed, everyone in the group wonders who else is a double agent (Epstein 1989: 116–17).

Just as using the lights to deceive a burglar receives daily feedback (no break-ins), intelligence agencies that run deception operations want to know whether the operation is working and, if not, how to refine the story. Therefore, it is crucial to have a source of feedback. Often a dispatched defector, also called a dangle, can fulfil that role as a mole; they have defected but are still working for the other side (Epstein 1989: 30). A mole or other source of feedback creates a deception loop with the deceiver feeding information to the target through channels the deceiver controls, while seeking to discredit or block rival channels. The target's response is then fed back to the deceiver by a mole, allowing the deceiver to modify ineffective parts of the deceptive message (Epstein 1989: 107–8, 118, 120). For example, the operation to deceive the Germans about the location, timing and size of the D-Day landings involved a deception loop. Captured double agents fed information to the target, the German High Command, who then unintentionally provided feedback via Ultra (the British ability to read coded German radio intercepts) to the British. The feedback allowed the deceivers, mainly MI6, to refine their message so that they convinced the Germans that, among other things, the Allies were not going to land at Normandy, so that there was no need to reinforce the landing area.

Ironically, German intelligence in general, and in France in particular, was good, especially their signals intelligence and infiltration of resistance groups. This efficiency, however, just made the Germans confident that they were not being deceived. To conduct an effective deception operation, the target must be confident of their own intelligence, so they will not believe that they can be duped, even when contrary information arises. This is crucial because there will always be contrary information. In 1944, for example, the Germans knew there had been reconnaissance flights over Normandy, just as there had been over many other possible invasion areas. The key was to use the German's confidence to reinforce the existing perception, held by Hitler and many top German generals that the landing would probably be in the Pas de Calais area, where the English Channel was at its narrowest, and discount information that pointed to Normandy (Epstein 1989: 107–8).

By reinforcing the target's perceptions, the deception continues because of a tacit symbiotic relationship between the deceiver and the deceived, who often have a powerful confluence of interests. Although symbiotic, this relationship strongly favors the deceiver, making it far more parasitic than symbiotic, although the symbiotic element is always present. A top Chinese agent, for example, appears to defect. He receives money and a channel to feed information or misinformation into US intelligence organizations. His CIA handler wants the defector to be important both to aid the United States and to further his own career. Furthermore, if the CIA handler is wrong about the defector's allegiance, then he is a dupe. Therefore, the CIA agent often will come to defend the veracity of his defector/agent. On top of this facet of their relationship, the defector and the handler work closely together and must develop a bond of trust in order to function. These factors make it extremely difficult to unmask a deception operation, since there are often ardent defenders of the validity of the deception within the deceived organization.

The symbiotic relationship between deceiver and deceived also applies to cases outside the world of spies. In 1935 the Anglo-German Naval Treaty limiting warship tonnages was signed. The German battleships *Tirpitz* and *Bismarck*, however, violated the treaty. The Royal Navy had evidence that the ships were too large, yet argued that the evidence was ambiguous. The problem, as a later British investigation determined, was that there was a notable "tendency of Naval officers and others who have taken part in negotiations to become advocates of the integrity of the persons of whom they secured the agreement" (Epstein 1989: 216–17). The Royal Navy wanted to trust the Germans or they would have had to explain why they negotiated the treaty in the first place with a government they did not trust to follow the terms of a treaty.

The same deceiver–deceived symbiotic relationship occurred during the Iran-Contra scandal. In the 1980s, the scandal involved the shipment of US arms to Iran in exchange for help freeing Western hostages in Lebanon. The CIA initially rejected Iranian feelers to initiate the operation, but the Iranians refined their story to appeal to fears that Iran would tilt toward the Soviets unless moderates in Iran won out. With the refined story, the Iranians found others in the Reagan administrations who were more receptive. The more the National Security Council in the form of Colonel Oliver North, Robert McFarlane, John Poindexter and others provided arms to Iran, the more they believed the operation was going to succeed. It never did. The US players continued to believe that they were dealing with a secretive moderate faction, not the Khomeini government itself, even when US intelligence reported that the shipments were being widely discussed within the Iranian government. Furthermore, none of the Western hostages were being freed as a result of the arms shipments. Final evidence that the Iranian government, not moderates, was running the deception operation is shown by the fact that after a Shiite newspaper exposed the operation, the "moderate" faction members who had been dealing with the United States were not arrested. In fact, they continued in their jobs. Even so, two weeks later at a television news conference, President Reagan still discussed dealing with Iranian "moderates" and his plans to "continue on this path" (Epstein 1989: 217–25). Ironically, the harder the target works to develop the relationship, the more likely the deceived is to believe the deception.

It is also easier to deceive an adversary because, as cognitive dissonance theory has found, in order to avoid emotional conflict or dissonance, humans tend to seek information that confirms and avoid information that refutes their beliefs (Myers 1996: 107). If humans did not have this tendency, it would be far more difficult to deceive an adversary. In foreign policy decision-making, this tendency sometimes leads to what Janis (1972) called groupthink, which occurs when the individuals in a group tend to come to hold the same beliefs and reach the same conclusions. This confirmation bias means that it is much easier for the deceiver if he reinforces a belief that the target already holds. The target will seek information to support that belief and tend to ignore or discount information that reinforces other conclusions. For the deceiver, this creates an ideal symbiotic relationship with the deceived.

Deception is almost always used with other means and, when it is, deception operations must be closely coordinated with military, diplomatic and economic measures. Therefore, all deception operations should be reported up the chain of command, so they can be coordinated. To maintain security, however, the lowest level does not need to know what deceptions are planned at the highest levels (Gerwehr and Glenn 2000: 23). In 1940, Wavell wrote, "In certain cases it may be essential that the commanders and troops executing a deception plan . . . understand both the real intention of the higher commander and the object of the deception plan" but "Normally, the fact that their moves . . . are being carried out purely to deceive the enemy, should not be disclosed to the troops concerned" (Brown 1975: 47). Security is fundamental to deception. If the enemy becomes aware that they are being deceived, it may spell disaster for the deceiver, although the enemy may still only know that they are being deceived. They probably still will not know which conclusion is valid, especially if the deception operation was based on valid information.

The Soviet deception planning for deploying nuclear missiles to Cuba in 1962 illustrates how extensive measures to ensure security should be for a deception operation to succeed. The Soviets detailed only five officers to work on the deception project and no secretaries were used to type final texts. To ensure security, everything was handwritten. Furthermore, no radio communications, coded or not, were sent about the operation. As a further cover, the plan was code-named Anadyr, the name of a river flowing into the Bering Sea, far from tropical Cuba. Troops were even told that they were heading for a northern destination, with some units receiving skis, felt boots and parkas. When the missile-laden Soviet ships passed through the Dardanelles, the Soviets lowered bundles of vodka, brandy, caviar and sausages to the approaching Turkish pilots so they would not board the ships and see the missiles and troops, who were kept below decks the entire voyage. The bribery worked. As one Soviet wrote, "Everyone likes to get presents, even pilots" (Hansen 2002: 53). Once in Cuba, the ships were unloaded mostly at night and ports were kept closed to foreigners, all to ensure secrecy, which was achieved. The United States only discovered the presence of the missiles after they were almost fully deployed (Hansen 2002: 49–53).

Far from ordinary people are required to conduct deception operations. Intelligence and creativity of a twisted, diabolical and almost criminal kind appears to be a primary prerequisite for effective deception officers. For example, a colleague wrote that "there was a diabolical mind operating" in London Controlling Section (LCS) Controller of Deception, Colonel John Henry Bevan, in charge of British worldwide deception operations during the Second World War, and it was said of Lieutenant Colonel Sir Ronald Evelyn Leslie Wingate, Bevan's deputy, that he could "think nine ways at once" (Brown 1975: 271).

Besides a highly intelligent, creative and diabolical mind, foreign experience is crucial to deception operations, whether gained in the government, military, on business or as a student. Wingate, for example, had vast foreign experience including time in Mesopotamia, Oman, Muscat and India. He also spoke French, Greek, Urdu and Arabic. Foreign experiences help deception planners get into the minds of their targets, so that they are familiar with the target's way of thinking and seeing the world.

During the Second World War, the British and Americans often recruited for deception operations creative people with backgrounds in theater and the arts who were expert at various techniques of illusion, including storytelling and acting, and often had a strong sense of humor (Holt 1994: 51–2). For example, a stage magician from a family of magicians, Jasper Maskelyne, was one of Britain's top camouflage experts. The LCS housed Wing Commander Dennis Wheatley, a leading British novelist and student of crime and black magic, while Douglas Fairbanks, Jr, a movie star, ran US Navy deception operations in the Mediterranean theater.

Surprisingly, a knowledge of history aids deception planners far more than a knowledge of military operations. The LCS, for example, did not rely on classics of military science, as would have been expected, but on lessons from history to devise their deceptions (Brown 1975: 274). History provides exposure to different ways of thinking based on different historical periods and, by so doing, increases empathy for other cultures, which is crucial to conducting deception operations. Therefore, effective deception officers do not require military experience. In fact, Allied military deception planning in the Second World War was largely conducted by amateurs. Bevan was a businessman-turned-colonel, yet helped coordinate British deception operations worldwide, while politician and future Senator Harold Stassen was Admiral Bull Halsey's deception officer in the Pacific. Peter Fleming, the older brother of James Bond creator, Ian, and an explorer-adventurer with extensive foreign experience, ran deception operations for Lord Mountbatten in the India/Burma theater (Bowyer 1980: 221). Dudley Clarke's "A" Force, the chief British deception organization in the Middle East and North Africa, included "a merchant banker, a chemist, a music-hall conjurer, a film scenario writer, an artist, intelligence men, [and] a don or two" (Brown 1975: 117), while the LCS was home to Harold Peteval, a millionaire soap manufacturer, Derrick Morley, a financier and shipowner, and Professor Edeward Neville de Costa Andrade, whose passion was "collecting old scientific books and useless knowledge." His "useless knowledge" helped him invent the little mechanical crickets used by Allied paratroopers on D-Day to identify friend from foe in the dark (Brown 1975: 272–3).

Even though military experience is not a requirement for deception officers, it is always key to have some experienced soldiers, such as Clarke and Wingate, among an organization's top deception planners. In fact, one of the most innovative deceptive minds in history resided in the skull of a soldier, Dudley Wrangel Clarke. Born in the Transvaal, South Africa in 1899, Clarke was raised in Britain. He joined the army at 16 and remained a soldier his entire life. An ideal staff officer, he never held command, but also never sought it. Field Marshall Alexander, his wartime commander, said that Clarke had "done as much to win the war as any other officer." Clarke was "clear thinking, blunt, respectful of authority, but never in awe of it. He had a deep appreciation of the flaws and foibles of himself and of others. He was the consummate realist." He also had a "puckish sense of humour [*sic*]" and "a boundless sense of the ridiculous." As an aide to General

Wavell for two years in Palestine in the late 1930s, he learned about military deception from Wavell, himself a dedicated student and practitioner of deception. In June 1940 Clarke was in London and wrote a memorandum that led to the founding of the Commandos, a name Clarke took from his childhood memories of the Boer commandos, even though Churchill later claimed it as his own creation. Late in 1940 Wavell summoned Colonel Clarke from London and for the rest of the war Clarke ran "A" Force, the senior deception coordinating staff in the Middle East. Clarke ran the first British double agents in the area and devised the idea of dummy military units, including creating whole phantom divisions, corps and armies. He coined the term "notional" for these bogus units; a term still used by military deception specialists today (Bowyer 1980: 221–3).

Even just one story will provide a hint about the depth of deviousness of the mind behind "A" Force. In North Africa at the end of 1942, Major Oliver Thynne learned that the Germans could distinguish flimsy British decoy aircraft from real planes because the dummies were supported by struts under their wings that were visible from the air. Thynne reported the development to Clarke. "Well, what have you done about it?" Clarke demanded. "Done about it, Dudley? What could I do about it?" Thynne asked. Clarke responded, "Tell them to put struts under the wings of all the real ones, of course!" (Bowyer 1980: 223). A straightforward mind would have ordered the dummies scrapped or fixed so that there would be no need to use struts. Clarke's devious mind immediately saw a way to take advantage of the flaw: by putting struts under real planes, German pilots, wishing to avoid wasting bombs on decoys, would not bomb the real aircraft. Furthermore, the Germans would then underestimate the number of real RAF planes and would face a devilish problem determining where the real planes were located amongst all the decoys, since they all now looked alike (Bowyer 1980: 221–3).

A mind like Clarke's is not a complete mystery. Individuals vary in their potential for deception and in the degree to which they expect deceit in others. Early studies in the 1970s by psychologists Richard Christie and Florence Geis sought to categorize people by whether they agreed with the statement, "There's a sucker born every minute." Unfortunately, almost everyone disagreed with the statement. Those who were low on deception measurements disagreed because they believed deceit (and therefore a sucker) was rare, while those high on deception measurements disagreed because they believed that there was *at least* one sucker born every minute.

The researchers persevered and developed a scale for what was called Machiavellianism, since the early tests included questions based on quotes from the Italian political theorist Niccolo Machiavelli's *The Prince* and *Discourses*. Sample quotes included "The best way to handle people is to tell them what they want to hear" and "The biggest difference between most criminals and other people is that criminals are stupid enough to get caught." There are no right or wrong answers and questions are scored on a five-point scale ranging from strongly agree to strongly disagree. The more you agree with the statements, the higher your Mach score.

The Machiavellianism test only reveals a potential to be deceitful, since an individual may believe that it is more effective to use deceit, yet choose not to employ it. High Machs, as they are called, are good at deception, although they will cooperate if they see that it will benefit them more than engaging in deception. High Machs can be either highly spontaneous, such as those who on the spur of the moment walk out on a restaurant tab, or highly structural, such as those who plan elaborate schemes to cheat on their taxes (Bowyer 1980: 198–201). High Machs, especially the highly structured kind, have the potential to be effective military deception officers.

Mach tests, however, do not determine who will be good at detecting deception, since, as Abraham Lincoln wrote, "You can fool all of the people some of the time" (Bowyer 1980: 75). Whether they are high or low Machs, everyone can be deceived. One exception are children, who are difficult and even sometimes impossible to deceive. Young children will not follow a fake or feint, because they perceive the world literally and only react to reality. For deception to be effective, the target must be smart enough to fashion the desired false reality. Therefore, if an individual is too inattentive, naïve or literal, they fail to create the false reality and are not deceived. Therefore, in one respect, the more intelligent the target, the easier they are to deceive, in that they will quickly follow the feint or misdirection. Scientists, intelligence experts, police officers, researchers, engineers and journalists are all highly trained in logical thinking. That very training in linear thinking, so useful in unraveling most scientific or intelligence issues, is totally useless, even dysfunctional when trying to detect deception.

The well-trained, intelligent, linear thinker, however, is not the very best target for deception. The best targets are those who are deceitful themselves; high Machs who use their deceptive potential regularly. As the seventeenth-century Cardinal de Retz said, "The most distrustful persons, are often the biggest dupes." As Bowyer (1980: 212) wrote, "Con artists feast on those with larceny in their hearts and devious counterspies prey on cheating spies." High Machs are good targets because those who are likely to be deceived need to have at least a sliver of the belief or idea that the deceiver wants to reinforce already present in their minds. Shakespeare's Iago, for example, is only able to deceive Othello because the Moor harbors some suspicion and jealousy already. If he did not, then Iago's hints about Desdemona's alleged infidelity would have found a distinct lack of fertile ground in which to fester and grow in Othello's mind. High Machs often feel that they are in control. They believe that they are on the lookout for, and therefore will detect, deception, even though it appears to do them little good. Stalin, for example, although a master of deception and deceit, was shocked in June 1941 when Germany invaded the Soviet Union. The high-Mach Stalin was certain that he was in control of the dangerous, deceit-filled game he was playing with Hitler. He clearly was not.

Even magicians, who are often high Machs and who know and practice deception or misdirection as a profession, can be deceived. In the 1920s, the American master magician Dr Harlan Tarbell and 11 colleagues attended a play, *The Charlatan*, starring Frederick Tilden as Count Cagliostro, an eighteenth-century alchemist. The magicians and the audience were amused by the many simple tricks used by Tilden in the play. Then the villain, a lawyer, forces Cagliostro to either back down in disgrace or accept a challenge to test his skills as a magician publicly. Cagliostro accepts and, in full view of the villain and a group of skeptics, displays a handful of sand, a clear glass flowerpot, a tall paper cone and a seed. The lawyer carefully inspects each prop, showing them to the audience to verify that they are as they seem to be and are otherwise empty. Cagliostro pours the sand into the pot, plants the seed in it, covers it with the cone and steps back. The lawyer then rudely intervenes by checking under the cone to verify that he has not been tricked. Then Cagliostro steps forward and confounds his enemy and the audience by raising the magic cone to reveal a full-grown rose bush. The audience applauded, but the magicians in the audience were amazed. They knew the trick; an old one called the "Indian Mango Tree," yet none of them could figure out how the trick had been done. Usually a confederate covertly passes the rose bush to the conjurer or the container is gimmicked to allow the rose bush to be hidden within it. The magicians thought some amazing new way of doing the trick had been devised. After the show, however, Tilden explained that he was just an actor and knew of no new way of performing the trick. He had simply had the lawyer, when he checked that the cone was empty, add the rose bush. By using the psychologically deceptive trick of using the hero's enemy in the play as the confederate, the magicians had not caught on that Cagliostro's archenemy had been his confederate. By buying into the play's story, the magicians had failed to detect the deception (Bowyer 1980: 397–8). A far from hopeful story for those attempting to avoid being deceived.

Even training in deception itself, whether in the fields of magic, the military or intelligence, usually does not aid in detecting deception. Unfortunately, as one author concluded from a study of deception in a broad range of human endeavors:

> There is no apparent defense against trickery, any more than the human eye can turn a movie into frozen frames or make the stick in the water look straight.... There seems to be little hope of detecting deception. The trained and innocent are duped.
>
> (Bowyer 1980: 226)

7

HOW TO DECEIVE

Stratagems

Deception is based on stratagems, which are defined in the *Oxford English Dictionary* as "An operation or act of generalship, usually an artifice or trick designed to outwit or surprise the enemy; a device or scheme for obtaining an advantage; cunning, used loosely for a deed of blood or violence." The British in the Second World War called these operations "special means," a:

> Vaguely sinister term that included a wide variety of surreptitious, sometimes murderous, always intricate operations of covert warfare designed to cloak overt military operations in secrecy and to mystify Hitler about the real intentions of the Allies. They were operations that, as M.R.D. Foot would observe, were "true to the tradition of English eccentricity; the sort of thing that Captain Hornblower or Mycroft Holmes in fiction, or Admiral Cochrane or Chinese Gordon in fact, would have gone in for had they been faced with a similar challenge; the sort of thing that looks odd at the time, and eminently sensible later."
>
> (Brown 1975: 2)

In fact, the British had a long history of such operations, causing Louis XIV's philosopher, Jacques Bénigne Bossuet, famously to exclaim, *"Ah! La perfide angleterre!"* (Brown 1975: 268).

British General Wavell counseled that the "elementary principle of all deception is to attract the enemy's attention to what you wish him to see and to distract his attention from what you do not wish him to see" (Brown 1975: 46). Therefore, there are two main categories of deception: hiding the real, such as with camouflage, and showing the false, such as with decoys. Several deception techniques, however, combine both hiding and showing. For example, in 1759 British Major General James Wolfe sent his troops up a goat trail to climb the "unscalable" Heights of Abraham to surprise the French in Quebec City. Wolfe hid the real (the route his army took), but also showed the false. He used French-speaking advance units, Highland Scots, who had been allied with France until the Clans' defeat on Culloden Moor in 1745, to deceive the French pickets.

Besides hiding and showing, deception operations can also be classified by whether they seek to increase ambiguity or to focus an adversary's attention on one possible option. The former type of deception occurred in 1973 when the Egyptians used more than 100 stratagems to increase ambiguity about whether, when and with what forces they would attack Israel. The latter type of deception occurred when British deception planners misled Hitler by reinforcing one option: the German high command's belief that the Allies would land at the Pas de Calais (Latimer 2001: 71).

Within the hiding/showing and ambiguity/misleading categories, the major deceptive stratagems are camouflage, decoys, mimicry, dazzle, disinformation, conditioning and changing rules or tactics. As the following examples illustrate, all of the major stratagems can be used at the tactical, operational or strategic levels of warfare, as well as in diplomacy and politics. Furthermore, altered images can be used as a basis for any of the seven stratagems, which can be employed in endless variations depending on the situation and the often boundless creativity of the deceiver.

Camouflage

Camouflage increases ambiguity by seeking to hide or mask an object, person or activity. Using natural (shrubs and bushes) or manmade (plaster, paint and netting) materials, electronic or thermal means, camouflage seeks to make personnel, objects or activity blend into the background like a polar bear in the snow so that an adversary will not detect the camouflaged personnel, object or activity. Camouflage relies, in part, on the principle of continuity, which predisposes people to perceive lines and colors as continuing in a natural progression. When an animal blends into the background, continuity predisposes us to see the background lines, shapes and colors as continuing through the camouflaged animal. Expectancy also plays a role. When we expect to see a farm and not an aircraft factory, we see a farm, not an aircraft factory. The key is to blend into the surroundings. As an ex-IRA man wrote, "You had to blend in with the local *population*: you had to look like a mechanic, or a postman, or a bank clerk, not Fidel Castro" (Collins 1997, cited in Gerwehr and Glenn 2000: 41; emphasis added).

Shakespeare provided an example of camouflage in *Macbeth* (V, IV, 6–9) when, much to Macbeth's dismay, Birnam Wood came to Dunsinane; "Let every soldier hew him down a bough, and bear it before him; thereby shall we shadow the numbers of our host, and make discovery err in report of us." Camouflage, however, goes much farther back than the bard. The Mongols raised dust clouds to camouflage the movement of their horsemen (Bowyer 1980: 29). In earlier times, however, camouflage was less used than today, since armies, for the most part, fought at arm's length or close to it. The key in Napoleonic warfare, for example, was not to hide, but to distinguish friendly from enemy troops. Therefore, bright, colorful uniforms were worn. It was not until the Second Boer War (1899–1902) and the widespread use of the rifle with its greater range that the British realized that their uniforms made them easy targets. They bid farewell to their red uniforms that had comprised the famous "thin red line" and adopted khaki uniforms to make them less conspicuous. The First World War saw the widespread use of camouflage for men, guns, aircraft, ships and supply dumps with the added new requirement that objects not only had to be camouflaged from enemies on the ground, but also in the air.

Camouflage reached new heights during the Second World War and was used to mask everything from explosives to airfields. In 1940, British would-be guerrillas were trained to use mines that were camouflaged to look like horse dung or lumps of coal (Breuer 1988: 43). The British also created fake bomb damage, relying on piles of debris stored near potential targets ready for use. The goal was to make a target appear ruined so the Germans would not bomb it again, even after it was repaired. At airfields, the British burned old tires to create smoke and dragged out ruined Spitfires and Hurricanes to make an airfield appear damaged beyond use. The British also painted craters on large canvas squares that were then bolted to airstrips. From the air, the airstrip appeared unusable and therefore not worth another German bomb. RAF pilots even sometimes thought the craters were real and refused to land. To maintain the deception, the "craters" would be removed several days later, by the time real craters would have been filled in (Breuer 1988: 68–9).

Camouflage could, however, be taken too far. In late June 1940, with Britain bracing for the Luftwaffe's air attack, the British government ordered that the tops of London's famed red, double-decker buses be painted battleship gray to make them harder to spot and bomb. If German bombers had been able to hit London buses their accuracy would have been nothing short of phenomenal, although their choice of targets would have left something to be desired (Breuer 1988: 22).

The German bombers faced a difficult task. They had to see their target from about 10 miles away at 19,000 feet. They had one unfailing aid to pinpoint British targets, however, the reflection of the moon off lakes, rivers, reservoirs and canals. Therefore, Charles Goodeve, leader of the British Directorate of Miscellaneous Weapons Development, was ordered "to hide the moon." A Directorate team tried using large nets supported by corks, but this only worked on ponds and small lakes. After much research, the team concocted a blend of coal dust and fuel oil that, when sprayed over a body of water, resulted in a dark, non-reflective coating. The Thames was chosen as a test site and even Churchill came to view the demonstration. Four launches started spraying the mixture on the river. The trial appeared to be a success, but it was not only the Thames that was being coated with the sooty, oily substance. The men on the launches were soon coated, making it difficult to see each other, let alone the river to navigate. Then the wind began to gust, the current increased and the non-reflective film broke up and scattered. Churchill, spattered with oil, left the scene in disgust. A naval craft was pressed into service to pick up dead dogs and cats from the river, while housewives downstream complained that the oily substance blowing off the river was ruining their washing hanging on outdoor lines.

Even with the disastrous test, the team did successfully hide the Coventry Canal, which the Germans were using as an orientation point. The camouflage was so effective that an elderly man and his dog, out for an evening stroll, fell in, having mistaken the canal for a new road. Even this, however, was an isolated success and, after much research, time and money, the plan "to hide the moon" was shelved. The moon would remain shining to aid the German bombers (Breuer 1988: 92–3).

Like the British, the Germans also used camouflage extensively. For example, after their rocket development site at Peenemunde was bombed in August 1943, the Germans rebuilt it, but left the craters where they were and heavily camouflaged the new buildings. The RAF believed that the site was still in ruins long after it was operational and did not bomb it again for nine months (Breuer 1988: 247). Just as the United States did when it used Hollywood talent to camouflage the huge Douglas Aircraft plant in Santa Monica, the Germans used set designers and artists from Italian motion picture studios to create effective camouflage that deceived the Americans when they landed at Anzio in 1944. The Germans' use of a decoy airfield made of wood in North Africa was somewhat less of a success. The British responded by sending a lone bomber over that dropped a wooden bomb (Brugioni 1999: 120–1).

Camouflage was also used in the secret war of spies. In December 1942, a German Abwehr agent, code-named Fritz, parachuted into England to blow up the de Havilland aircraft works at Hatfield, just north of London. On 27 January 1943 he radioed Paris: "Will attempt sabotage this evening at six o'clock." Just before dawn the next day, the Abwehr received the signal, "Mission successfully accomplished. Powerhouse blown up." Captain Stephan von Grunen, chief of the Paris Abwehr station, was skeptical that one man could damage what must have been a well-guarded plant. To verify Fritz's report, the Germans flew two reconnaissance flights over the plant. The photographs showed gaping holes in the roof of the aircraft factory's powerhouse that revealed scattered bits and pieces of destroyed generators.

Fritz then radioed that British counterintelligence units were closing in on him. He slipped aboard a freighter to Lisbon and returned to Paris, a hero of the Third Reich. He was even treated to a celebratory dinner at Maxim's. Fritz, however, was a double agent, dubbed Zigzag by the British XX Committee, the doublecross committee that ran double agents. He was actually Eddie Chapman, a British criminal with a penchant for safecracking. Jailed in the Channel Islands, he was released when the Germans invaded. He volunteered to spy for the Nazis and was sent back to England, where he promptly turned himself in and became a trusted British spy. Major Jasper Maskelyne, an accomplished magician in civilian life, created the fake bomb damage at the de Havilland plant. Covering the roof of the power plant with canvas, his team painted the damage on the canvas as it should have appeared if Fritz had been successful. Papier-mâché was used to create what appeared to be bits and pieces of destroyed generator. When combined with chunks of brick and cement blocks, RAF photo-interpreters, shown aerial photographs of the damage, could not tell that it was fake. It also fooled the Germans and established Fritz-Zigzag-Eddie as a bona fide German agent (Breuer 1988: 197–9).

Camouflage still works in the era of thermal sensors, satellites and reconnaissance aircraft. Early in the 1982 Falklands War, the British bombed the airstrip at Port Stanley with Vulcan bombers flying from Britain on marathon, trans-Atlantic flights. After the raid, aerial reconnaissance photographs showed craters on the runway, rendering it inoperable. It was only after Royal Marine and British Army units had landed at Port San Carlos and "yomped," as the troops called it, across the island, that they heard strange sounds at night. They soon realized that the strange sounds were Argentine C-130s operating from the airstrip. Almost all of the Vulcans' bombs had missed. After the British raid, however, the Argentines had camouflaged the airstrip to make it appear to be cratered, but had continued to use it at night.

Camouflage can also take a more brutal turn. When the US Marines were in Beirut in 1983, their adversaries used women to reconnoiter Marine positions. One such scout appeared to be "a heavyset middle-aged woman – or a large man dressed in a woman's clothing." A Marine finally got tired of the constant scouting and shot the "woman" with his M-16. When the woman fell, an Amal gunman who had been hiding behind the woman and using her as camouflage scrambled into a nearby building (Gerwehr and Glenn 2003: 38).

Today, military forces use camouflage at sea, in the air and on land. Warships are painted gray, while aircraft are either gray or blue and white. Nuclear submarines use the vastness of the ocean as camouflage. The Soviets took hiding their submarines one step farther, however, when they planned to send a submarine to Cuba. They suspected that the Americans were detecting their submarines, which were far noisier than US submarines, as they left port, so the Soviets camouflaged their submarine by having her leave port directly under a merchant ship. The screws of the merchant ship made so much noise that the Americans could not detect the Soviet submarine cruising beneath her. Stealth aircraft, such as the F117A Nighthawk and the B-2 Spirit bomber, as well as the American stealth ship, the *Sea Shadow*, employ a range of camouflage. Their black color makes them more difficult to see at night. They have no sharp surfaces that can reflect radar and their exhaust outlets are hidden to thwart thermal sensors. On land, camouflage is also still employed. In 2003, Iraqi Fedayeen wore civilian clothes and drove civilian vehicles to attempt to get close to US forces. As Mendez (2000), and Mendez and Mendez (2002) demonstrate, spies also often use disguises, which is really just personal camouflage.

Camouflage, however, sometimes conflicts with other issues. In the 1980s a proposal to paint over the bright tail-fin unit designations on US Navy aircraft, especially the F-14 fighters, with camouflage gray raised such morale and pilot retention concerns that the bright emblems were retained despite their eye-catching properties.

Decoys

Whereas camouflage seeks to hide a real target, a decoy is designed to show a false target, which the deceiver wants his enemy to attack. Decoys are an ancient idea.

During the siege of Syracuse in 212 BC, the Greek defenders built a false beach to decoy the Romans. The initial Roman assault landed on the decoy beach and, as the Romans floundered in the surf, the Greeks slaughtered them (Bowyer 1980: 21–2).

In modern times, decoys were used extensively during the First World War. At the start of the war, Churchill planned to create a decoy fleet to lure German U-boats into traps. Merchant vessels would be disguised as battleships. The beauty of the plan was that even if the enemy learned about the dummy fleet, they would never be certain whether a battleship target was real or a decoy. This feature is true of all decoys, since even knowing that decoys exist does little to simplify the attacker's problem of distinguishing the real from the decoy. In fact, even just a rumor that decoys exist, when none in fact do, can make the attacker's task far more difficult and uncertain.

In October 1914, ten old passenger and cargo ships were given broader decks, greater length and warship-like bows. Smoke generators were installed to belch smoke. The ships were then ballasted with 9,000 tons of stone to ride low in the water to mimic a warship's low profile so they could serve as decoys for ten battleships (Massie 2003: 141–3). The decoy fleet was at first anchored in the outer roads at the Grand Fleet's main base at Scapa Flow to draw the torpedoes of any U-boats that came calling, but in March 1915 the decoy fleet was sent to the Mediterranean to support Allied operations at the Dardanelles. On the night of 30 May 1915 in the Aegean Sea, a U-boat sighted the former S.S. *Merion*, a 12,000-ton liner, which had been converted to mimic the 30,000-ton *Tiger*, the Royal Navy's newest and largest battlecruiser. The U-boat torpedoed her. The 70-year-old captain and all but four of the 117-man crew abandoned ship as the ex-*Merion*, ballasted with rock, slowly sank on an even keel. "*Tiger*'s" cover was probably blown when her supposed 100-ton turrets were left floating in the water above the ship's grave (Bowyer 1980: 76–8). It is unknown how much confusion the decoy fleet caused the Germans, but the fleet cost the British about a million pounds, four British seamen their lives on the ex-*Merion*, and cost Germany atleast one torpedo (Massie 2003: 306–8).

Decoys were also commonly used during the Second World War. During Operation Torch, the Allied landings in North Africa, the British spread disinformation through double agents to lure U-boats away from the Torch troop convoys onto an empty convoy, SL125, heading to England from Sierra Leone. The convoy paid the price of being a decoy; U-boats sank 13 ships. The ships, however, were empty. The convoy did its job and its commander, Rear Admiral C.N. Reyne was even "congratulated for losing ships." The vulnerable and valuable Torch troop convoys, however, did not lose a man (Brown 1975: 233–4).

During the Battle of Britain and the Blitz, the British created decoy airfields to lure Luftwaffe attacks away from real RAF airfields. Colonel John F. Turner, in command of the operation, set up two rows of flares or lights in open country two miles from several RAF airstrips, hoping the Germans would think they were emergency airstrips. The Germans did and promptly bombed the decoy airstrips.

Deceptions must evolve, and Turner's "airstrips" evolved. He dimmed the lights to simulate an airstrip operating at night and added phony recognition beacons. When German planes were heard overhead, men ran across the fake "airstrips" with automobile lights mounted on wheels to simulate RAF planes taking off and landing. The fake planes attracted many German bombs. Sometimes deception was used in reverse. One night, lights at a decoy RAF airfield were dimmed, while the lights at a real field were fully lit. The Germans determinedly bombed the dimmed decoy field (Breuer 1988: 60–8). During one period, the decoy airfields attracted 440 air raids compared with only 430 against genuine RAF airfields. The decoys, however, could be overvalued, especially by their sometimes possessive creators. In one case, a Flight Sergeant pleaded with his commander as the Germans attacked his decoy airfield, "We need fighter cover! They're wrecking my best decoys" (Latimer 2001: 87–8).

The use of decoys, of course, did not end in 1945. During the Cold War, US analysts discovered several new submarines in photographs of the Soviet's Northern Fleet base at Polyarnyi. Having had no hints that any new submarines were joining the fleet, the analysts were appalled that the Soviets could build and deploy submarines in such numbers so quickly and secretly. Then, after a severe winter storm, US satellite photographs showed several of the new boats bent in half at right angles. The Soviets had deceived the Americans by using rubber decoy submarines (Cockburn 1984: 448).

Today the use of decoys is widespread. Intercontinental ballistic missiles (ICBMs) have dozens of multiple, independently targeted, reentry vehicles (MIRVs) that act as decoys. Canadian CF-18s and US A-10 Thunderbolt II "Warthogs" have false cockpits painted on their undersides, to give the pilot a fraction of a second advantage in dogfights (Gerwehr and Glenn 2003: 12–13). During the Second World War, chaff was invented. Strips of aluminum foil, chaff causes a mass of static on radar screens. It is still used today by ships and aircraft to present a decoy target to incoming radar-guided missiles. During the Falkland's War, the British used a more dangerous form of decoy. RN helicopter pilots flew their craft just above some of the ships of the task force, especially the valuable carriers, acting as decoys to draw off incoming Exocet missiles. The Argentines used simpler and much safer decoys for their guns. British Major Gerald R. Akhust, Royal Artillery, later wrote, "I fired lots of rounds at suspected gun positions only to find out later it was two old tires and a drain pipe draped with a camouflage net. At deception, the enemy was very good" (Mullen 2005: Online).

Decoys continue to evolve and improve. The United States has a full panoply of decoys, including M-1A1 Abrams tanks, Humvees, self-propelled guns and aircraft, such as F-15s and F-16s. The new inflatable decoys often have generators to keep them fully inflated, so there will be no telltale creases. Decoy airstrips now have oil stains and aircraft tire tracks to make them more realistic (Brugioni 1999: 120–1). During the Gulf War, the Iraqis used decoy tanks they had purchased from an Italian company (Latimer 2001: 260), while during the fighting in Iraq in 2004, US troops used simple decoys that consisted of a helmet, sunglasses and a kerchief atop a stick to draw sniper fire (Sanders and Chu 2004: A14).

As with camouflage, even in the age of satellites, thermal imagers and Unmanned Aerial Vehicles (UAVs), decoys are effective. During NATO air operations in Kosovo in the 1990s, the Serbs used decoy tanks, armored personnel carriers (APCs) and guns. The decoys were made of wood, chicken wire, black logs and old truck tires. Some also had charcoal braziers to provide appropriate heat signatures as if from a warm engine or a gas can that made the "tank" explode convincingly when hit. A decoy bridge, which NATO destroyed several times, was made of polyethylene sheeting. It protected a real bridge 300 meters away. Another decoy bridge at Kosovo Polje was made of timber and roofing felt, which gave off the same thermal signature as tarmac and made a convincing roadway.

Apart from making attacks far less effective, decoys make battlefield damage assessment (BDA) extremely difficult. A Pentagon report published after the NATO air campaign in Kosovo found that of 744 "confirmed" hits during the conflict, only 58 air strikes had actually found their target. NATO also erred in its BDA. It claimed to have destroyed about 120 tanks, 220 APCs and 450 artillery pieces and mortars in the 78-day bombing campaign. In reality, thanks in large part to decoys, NATO only destroyed 14 tanks, 18 APCs and 20 guns and mortars (Latimer 2001: 308). Using more than one type of sensor, however, can help distinguish real targets from decoys. If Operation Allied Force, for example, had used ground forces as well as aircraft, NATO would have been far less fooled by Serb decoys that appeared to be military vehicles from 20,000 feet, but were clearly just wood and sheet metal when seen from nearby hills (Gerwehr and Glenn 2003: 22).

Surprisingly, although better sensors may lead to the discovery of more camouflaged vehicles, they also often lead to the discovery of more decoys. Worse, improved sensors often provide little or no aid in distinguishing between real targets and decoys (Gerwehr and Glenn 2003: 60, 63). Contrary to popular belief, modern satellites are little better at reconnaissance than Second World War photoreconnaissance aircraft. Although satellites today have far better cameras, they orbit 100–330 miles up. Planes in the Second World War, which had less precise cameras, were a mile or less, and sometimes even only a few hundred feet above the ground. For example, taking pictures from an altitude of 100 miles with a telescope that has a focal length of 64 feet (the maximum that will fit into US KH-11 satellites) in perfect conditions permits observation of objects down to 6 inches, which is about the same resolution as aerial cameras of the 1940s. Therefore, the requirements to create decoys and camouflage objects today are not much more challenging than they were in the 1940s, when decoys and camouflage routinely duped aerial reconnaissance analysts. Granted, there are new types of sensors today, such as thermal and electronic, but they can also be fooled. A brazier can mimic the thermal signature of a tank and a panoply of electronic countermeasures has been created to fool electronic sensors. Furthermore, most US adversaries know when US photographic and signal satellites, which monitor electronic signals, are within range. This knowledge allows adversaries to take deceptive measures when satellites are overhead. Therefore, in some ways, satellites are easier to deceive than the Second World War-era photoreconnaissance aircraft, which could appear at almost any time (Epstein 1989: 158–60; Latimer 2001: 309).

Mimicry

Deception almost certainly goes back as far as warfare itself, but the first recorded mention of deception in warfare was a case of mimicry. In 1450 BC, after a failed assault on the city of Jaffa, the Egyptian General Thot's army retreated, leaving behind baskets containing "surrender gifts" for the city's victorious defenders. When the defenders greedily hauled the baskets into the walled city, they were set upon by a select force of Thot's soldiers who had been hiding in the baskets (Bowyer 1980: 17). Mimicry seeks to make one thing appear to be something else. The soldiers in the baskets appeared desirable by mimicking surrender gifts, but mimicry can be of a larger force, a smaller force, an enemy unit, a benign object, a leader, a retreat or an attack.

Gideon, the Old Testament judge of the Israelites, mimicked a larger force when he attacked a superior Midianite army. Gideon's army attacked at night and blew many more battle trumpets than the size of his army warranted. Each trumpet signified another non-existent unit. Fearing that they were facing a much larger force, the Midianites fled (Bowyer 1980: 20). In February 1779, George Rogers Clark used a similar form of mimicry when he approached Vincennes, Indiana, with only 47 men but with far more flags than his small force warranted. His mimicry convinced the 150 British soldiers and their numerous Indian allies in Vincennes that they faced a far stronger force than they in fact did. The Indians fled and the British surrendered to Clark's numerically inferior force (Holt 1994: 51).

Such mimicry of larger forces also occurred during the US Civil War. Amazingly, Confederate General John Bankhead Magruder's 8,000 men kept General George B. McClellan's Union Army of 120,000 men at bay on the peninsula below Richmond in 1862 for more than two months through the use of mimicry. Quaker guns (stripped logs, painted black, with wagon wheels tacked onto the sides), scarecrows and repeatedly marching the same units through clearings within sight of the Federal lines allowed Magruder's tiny force to mimic an army 10 times its real size. McGruder's theatrical flair reinforced McClellan's firm belief that he faced an enemy host of at least 85,000 men, earning McClellan the nickname, "The Virginia Creeper" for his tardy advance toward Richmond (Latimer 2001: 26–9). Unbelievably, by the end of the campaign McClellan believed that he faced a force that was larger than his own (Gerwehr and Glenn 2000: 33; Holt 1994: 51). Quaker guns, however, did not die with the Confederacy. At the start of the First World War, an old Turkish battleship, which lacked real guns, mounted wooden guns to appear fully armed when she was not (Massie 2003: 23).

One of the finest examples of mimicry occurred in the Second World War, when the British and Americans used a handful of personnel to mimic two entire armies. The British knew that the Germans thought the Allies would land at the Pas de Calais and, therefore, undertook Operation Fortitude to reinforce this German belief. To mimic a threat to the Pas de Calais, the British and Americans fabricated an army of more than half a million men; the First United States Army Group (FUSAG) in East Anglia and Kent. Supported by double agents who fed the Germans reports that more US troops were arriving in England than in fact were reaching the sceptered isle, movie set designers from Shepperton Studios created the fictitious FUSAG force with what appeared to be armadas of landing craft, squadrons of tanks, batteries of artillery and rows of trucks; all made of inflatable rubber (Brown 1975: 603). Lumber, wire and canvas were used to create barracks and warehouses for the "soldiers" and "supplies" massing in East Anglia, across from the Pas de Calais, while hundreds of canvas and wood landing craft, called Bigbobs, floated on empty oil drums in harbors all along the coast (Brown 1975: 91–3). Some of the facilities were purposefully poorly camouflaged so the Luftwaffe would detect them. Studio set designers even built an enormous fake oil-storage complex near Dover to support the phantom FUSAG. The press covered visits by King George, General Bernard Law Montgomery and Eisenhower to the site. Bogus radio traffic filled the air and Lieutenant General George S. Patton, Jr took command of FUSAG, since the Germans expected Patton, the Allies' toughest, hardest-charging and, the German's believed, best general to command the invasion. Patton was upset he would not be leading the real invasion, but he relished the act, telling aides, "I'm a goddamned natural-born ham!" (Breuer 1988: 235).

All effective deception operations use multiple stratagems. Beyond using camouflage and decoys, the FUSAG mimicry ruse also included disinformation. When General Hans Cramer, last commander of the Afrika Corps, was repatriated home from a POW camp in Wales for health reasons in May 1944, he was driven to London by two agents of the XX (Double Cross) Committee. They drove right through the assembly areas in southwest England for the invasion of Normandy. Cramer, however, was told that they were driving through the Dover area. Since signposts had been taken down in 1940 to hinder what was feared to be an imminent German invasion, Cramer dutifully reported to Hitler upon his return to Berlin that the Allies had a huge army massed around Dover for a landing on the Pas de Calais (Breuer 1988: 236–7).

The other fictitious army, the British Fourth, was created to mimic an army preparing to invade Norway from eastern Scotland. The plan was designed to pin down 27 German divisions in Scandinavia before and just after D-Day. Unlike FUSAG's bogus tanks, landing ships, barracks and oil-storage facilities, only a handful of overage and unfit soldiers supported by bogus radio traffic mimicked the phantom Fourth. As with Patton for FUSAG, the Germans knew and respected the commander of the Fourth Army: General Andrew "Bulgy" Thorne, who had been a military attaché in Berlin before the war (Breuer 1988: 238; Brown 1975: 603).

Such mimicry of a larger force continued after the Second World War. At the 1955 Tushino Air Show, for example, the Soviets flew a squadron of Bison bombers over the airfield. The bombers cruised out of sight and then another squadron roared over the spectators. Squadron after squadron flew overhead, convincing Western air attaches that the Soviets had a fleet of Bisons, which helped give birth to the bomber gap. The Soviets, however, had simply flown the same squadron over the airfield multiple times (Cockburn 1984: 446). More recently, in 1990, the US Army had the ability to mimic a major headquarters to make it appear that the 5th Corps headquarters was moving to Turkey to prepare to launch a major attack from Turkey into Iraq (Clancy *et al.* 2004: 193).

Mimicry can also involve a disguise that makes a warship appear to belong to an enemy navy. Such a disguise can be as minimal as a false flag. In August 1914, the German battle cruiser, SMS *Goeben*, flew a Russian flag as it approached the French North African port of Philippeville. After using the flag of France's ally to get within range, the *Goeben* ran up the German flag and shelled the surprised occupants of the harbor (Massie 2003: 34). Mimicry can even be accomplished with no flag at all. In September 1914, Admiral Otto von Spee led his German East Asia Squadron into Bora Bora in the French Society Islands. The Germans flew no national flag and the French assumed the ships were British. The French authorities came out to visit the "British admiral" and met only English-speaking German officers. The French brought bouquets of flowers, war news and even provided information about the islands' defenses. The Germans paid for supplies with gold. As they departed, the French unfurled a large tricolor in a farewell salute. The Germans returned the salute by unfurling the German naval ensign (Massie 2003: 192).

Sometimes such mimicry of an enemy ship requires more than just a false flag. In March 1942, the British used an old US destroyer, the *Buchanan*, which they renamed the HMS *Campbelltown*, to blow up a huge dry dock in St Nazairre, France. They significantly altered the *Campbelltown*'s appearance so that it mimicked a German sub-chaser. The ship was then loaded with 5 tons of explosives and, through a daring raid that relied heavily on the fact that the Germans did not recognize the ship as a threat until it was too late, ran the ship into the dry dock and destroyed it. The loss of the dry dock deprived the Germans of one of the few places where they could repair their big surface ships, such as the *Bismarck* and *Tirpitz* (Breuer 1988: 121–3).

Such mimicry of enemy forces can also occur on land. During the Second World War, the Germans used the Brandenburg Regiment to mimic Polish troops during the 1939 invasion of Poland. The same unit mimicked Dutch soldiers so that they could seize key bridges during the 1940 invasion of the Low Countries, and Soviet NKVD personnel during the 1941 invasion of the Soviet Union. Finally, in 1944, the unit mimicked Americans to sow confusion behind the Allied lines during the Battle of the Bulge (Latimer 2001: 89–92). During the Korean War, Chinese troops often mimicked Republic of Korea troops, such as when they used South Korean uniforms to approach US positions near Unsan in November 1950 (Spurr 1988: 144).

Soldiers can also mimic voices to deceive the enemy by issuing false orders over the radio. Such measures may not withstand scrutiny, but in a crisis or combat, such approximations are often good enough to deceive long enough for an adversary to do what the deceiver wants. In the 1990s, the Chechens, for example, listened on the radio for Russian units in trouble and, when they asked for instructions, the Chechens mimicked Russian commanders and ordered them to retreat (Gerwehr and Glenn 2000: 21, 27, 33).

Spies often mimic adversaries to gather information or to become double agents, while government forces often mimic rebels or guerillas to pose as agent provocateurs. As Henderson and Goodhart (1958) chronicled, the British, for example, had some of their own troops mimic Mau Mau guerillas in Kenya in the 1950s. Mimicking adversaries is even easier online. In 1997, the Tamil Tigers hacked into Sheffield University's computer system. The Tamils captured the legitimate user IDs and passwords of respected academics in order to use their e-mail accounts to ask people to send money to charities in Sri Lanka that supported the Tamil's cause (Vatis 2001: 4).

Mimicry is also used to make a force appear weaker or benign. In the fall of 1915, the Royal Navy converted three tramp steamers into auxiliary cruisers or Q ships. They were armed for anti-submarine warfare but mimicked the appearance of benign unarmed merchant ships. U-boats carried few torpedoes, so after launching one torpedo they often surfaced to finish off a victim with their deck gun. The Q ships had guns camouflaged in collapsible deckhouses or lifeboats. A duplicate command post was also hidden as a coil of rope, allowing the captain to remain hidden while he directed an action. Pipes were even added so that steam could be pumped out to make it appear as if the ship's boilers had ruptured. Small explosive canisters were set around the deck, so if a U-boat opened fire with her deck gun, the fake explosions would confuse the gunners' aim. To ensure survivability, extra watertight compartments were built below decks and huge beams of Canadian spruce or empty barrels were packed into the hold. The crews even dressed in mixed sailors' garb bought from secondhand stores, never saluted and, in one case, included a sailor dressed as the captain's wife so as to mimic better a civilian crew. The British converted 193 ships into Q ships during the war and they sank 15 U-boats, with four more "possible" but unconfirmed sinkings (Bowyer 1980: 79–80; Massie 2003: 718–25; Smith 2005: Online).

During both world wars, German raiders mimicked merchant ships, so as to appear non-threatening to real merchant ships. The mimicry was effective. Between September 1939 and March 1943, for example, commerce raiders sank 129 Allied ships totaling 800,310 tons, even while they disrupted trade and tied down naval resources protecting convoys and hunting the raiders (Hinsley 1981: 167). German mimicry, however, did not always work. During the First World War, they used a Hamburg-America Line excursion steamer, the *Königin Luise*, to mimic a British Great Eastern Railway steamer when she went to sow mines off the mouth of the Thames. The ship had just started laying mines when the British light cruiser *Amphion* saw through the mimicked disguise and sank her, rescuing 56 of the crew of 130. The prisoners were put in the light cruiser's bow, since, as an RN officer said, "if we go up on a mine, they might as well go first." The *Amphion* did later hit a mine, killing all but one of the prisoners (Massie 2003: 77–8).

Mimicry of merchant ships is still done today. In 2003, for example, the Japanese raised a North Korean spy ship that three Japanese coast guard ships had sunk on 22 December 2001. The ship had changed its color scheme many times to mimic Chinese and Japanese fishing trawlers. It even carried pre-cut paint sponges in the shapes of Chinese and Japanese characters so the crew could quickly rename their ship and change other identifying marks. Hidden in her hull were four giant Russian-made engines that could propel the ship at up to 33 knots, far faster than any ordinary fishing boat. Two large stern doors hid a cavity that contained a speedboat, which was used to land agents on the Japanese coast. Spy ships are supposed to keep a low profile, but this one failed. Once it was raised and towed into Tokyo Bay, nearly a million people were expected to tour the ship during its first summer as a tourist attraction (Magnier 2003: A4).

Mimicry to appear benign also occurs at the individual level. In 1968, Marines fighting in Hue, Vietnam battled VC units in a hospital disguised as patients. One Marine also encountered a person wearing a black habit, whom he assumed was a nun. Fortunately, the "nun's" pistol misfired and the Marine's life was spared. The nun was a VC soldier (Gerwehr and Glenn 2000: 47). The 1991 IRA mortar attack on No. 10 Downing Street was launched from a civilian van concealing a Mark 10 mortar (actually three oxyacetylene cylinders on a rack) fired through an opening cut in the roof. The van mimicked a thousand other civilian vehicles on the streets of Westminster and was parked 200 yards from the Prime Minister's residence, the driver escaping unnoticed on the back of a waiting motorcycle (Gerwehr and Glenn 2000: 47).

Mimicry can also occur in politics. In 2005, commentator Armstrong Williams mimicked an independent voice when, in fact, the Bush administration was found to have broken the law when it used taxpayer funds to pay the media pundit $240,000 to promote the president's education policy (Hamburger *et al.* 2005: A1). Democrats also charged California Governor Arnold Schwarzeneggar with using taxpayer money to produce mock news videos to promote the Governor's policies. Schwarzeneggar's spokesman said the spots were just video press releases, but they featured an ex-reporter and included suggested introductory remarks for anchors, which made them mimic news reports (Morain 2005: B1, B11).

One of the more audacious cases of mimicry of a leader started on 14 March 1944, when Lieutenant Colonel J.V.B. Jervis-Reid happened to see a photograph of Lieutenant Meyrick Edward Clifton James of the Royal Army Pay Corps. James was a minor stage actor in peacetime but bore a striking resemblance to Montgomery. Their stance, build and walk were almost identical, and James could imitate Montgomery's "rapid, somewhat squeaky diction without trouble." There were significant differences, however, including the fact that James suffered from alcoholism, while Montgomery was a teetotaler. The plan was for James to mimic Montgomery, and parade through Gibraltar and Algiers just before D-Day to make the Germans believe that, with Montgomery outside Britain, the invasion would not be launched in early June. Separated from his wife and the bottle, James was told that he was impersonating Montgomery for a training film. Only after Monty insisted that if James was going to impersonate him, he must be paid like him, was James told his true mission. James boarded a Liberator and flew to Gibraltar, but had hidden a flask of gin aboard. Stricken by stage fright, he was soon drunk. The operation appeared to be headed for disaster, but his handlers were men of action. They held James' head out an open window into the icy wind, then forced him to vomit, before walking him up and down the aircraft as they massaged, slapped and doused him with cold water. By the time they landed, James was sober.

The Governor, who knew enough about the plan to play his part, met "Monty" and the general was introduced to two Spanish bankers and saluted by some troops. At Algiers "Monty" reviewed a French guard of honor and met with the Supreme Commander, General Sir Henry Maitland Wilson. Then stories surfaced that Montgomery had been seen staggering around the streets of Algiers with a cigar in his mouth. Montgomery did not drink or smoke. That ended the operation. James/Monty vanished and was never seen again.

After the war, James told his story in a book and movie, *I Was Monty's Double*. He claimed great success for the operation, although there was nothing found in Abwehr documents after the war to show that the Germans had taken into their calculations the fact that "Montgomery" had visited the Mediterranean theater in early June 1944. James' performance, it appeared, had been "to an empty house" (Brown 1975: 608–12). Even mimics need an audience.

Two general types of activity are usually mimicked; retreat and attack. In about 1200 BC Joshua captured the city of Ai by staging a feigned retreat. He concealed most of his force out of sight behind Ai, and then marched the remainder of his army up to the city to offer battle. When the men of Ai came out to fight, battle was joined and Joshua's forces fell back, mimicking a rout. The men of Ai rushed after their defeated foes. Joshua's hidden force then appeared and attacked the city, overwhelming the small force left within to defend it. The men of Ai were now trapped between Joshua's forces in the city and those outside, and were slaughtered (Bowyer 1980: 17; Gerwehr and Glenn 2000: 32). In 1066 at Hastings, William's troops also mimicked flight to entice the English line off its strong defensive position atop a hill. The result: William won, lost the nickname "the Bastard," gained the nickname "the Conqueror," and England became Norman (Latimer 2001: 10).

Attacks can also be mimicked, either as feints or demonstrations. Feints are conducted near where a real attack will occur and may come into contact with the enemy. Demonstrations are conducted in areas where no real attack is planned and do not come into contact with the enemy. The British, for example, staged a demonstration on the morning of 23 October 1942, to distract the Germans from their attack that initiated the Second Battle of El Alamein. The British staged a mock invasion on the coast behind the German lines complete with motor torpedo boats, smoke, the noise of anchor chains rattling and men shouting orders over loudhailers. There was even the smell of engines to add to the effect (Brown 1975: 126). During Operation Desert Shield (1990–1), the Coalition used mimicry supported by disinformation and decoys to create a demonstration. The goal was to convince Iraq that the Coalition attack would come through central Kuwait and not out of the western desert. The Coalition broadcast tank noises over loudspeakers, positioned dummy tanks and artillery, and simulated radio traffic to make the Iraqis believe that units were still in their old locations near the coast when, in fact, they had moved far inland to execute the Hail Mary left hook attack across the desert (Gerwehr and Glenn 2000: 34). The United States used demonstrations and feints during the 2003 war in Iraq, such as those executed by the Third Infantry Division at Karbala and the Marines south of Baghdad. In 2004, the Marines also mounted a feint when they positioned armored vehicles south and southeast of Fallouja as if they were preparing to attack the rebel-held city. The actual attack came from the north (McDonnell 2004: A9).

A mimicked attack can also be staged against your own side. At the start of the Second World War in Europe, the Germans mimicked a Polish attack on a German radio station near the border by staging Operation Himmler. The goal was to make it appear that the Poles had started the war. On the night of 31 August 1939, SS troops, dressed as Polish soldiers, attacked a radio station near Gleiwitz. A Polish-speaking German broadcast anti-German propaganda for a brief time and then the SS troops departed. They even left behind 12 dead "Polish" soldiers, actually German criminals (code-named "canned goods") who had been drugged, brought to the station and then shot dead. German reporters were rushed to the site the next morning and photographs of dead "Poles" appeared in newspapers across Germany. The timing of the operation, however, was off. Only nine hours after Operation Himmler, Germany invaded Poland. Few, if any, believed that such a large-scale attack, which would have required extensive and lengthy preparation, was in retaliation for the "Polish" Gleiwitz raid (Breuer 1988: 5–8).

Dazzle

Dazzle, a type of showing, involves increasing ambiguity by overloading an adversary's decision-making process or sensors with unimportant information or noise. Background noise makes intelligence gathering and decision-making harder to begin with, so deliberately making background noise increases ambiguity and makes perceiving reality that much more difficult. False bow waves, which were painted on many warships during the world wars to confuse an attacker about how fast the ship was traveling, is one form of dazzle. The Germans in the First World War used dazzle by giving their reserve units the same numerical designation as their operational units. Therefore, the same unit appeared to be in two places at once. Dazzle can be electronic, such as chaff, which overloads radar by producing a mass of images, or jamming, which overloads radios with meaningless signals. It also can be an artifact of a society. During the Cold War, some Soviet spies complained that the United States provided too much open information, even while American spies made the same complaint about the volume of Soviet open sources (Mercado 2005: 46). So much information was published by both superpowers about national security and weapon systems that it was extremely difficult to determine what was important.

Activities can also be masked by dazzle. Before the 1944 Normandy landings, the Allies flew photoreconnaissance flights over possible landing beaches from Norway to the Balkans, as well as those they needed to fly over Normandy. Therefore, the Germans received a mass of information about reconnaissance flights over northern, western and southern Europe. They faced the near impossible task of determining which flights presaged an invasion and which were dazzle. The problem was that all of the flights, both real and dazzle, looked exactly the same.

Dazzle is especially useful if an adversary already knows the deceiver's plans and the deceiver wants to confuse the issue. For example, during the 1944–5 German winter offensive in the Ardennes that led to the Battle of the Bulge, the Germans knew that Patton's Third Army would move to attempt to relieve the encircled "battered bastards of Bastogne": the US 101st Airborne Division. Patton's dazzle-ruse was to issue five sets of orders over the radio to his army. Only his own officers knew which set was correct, so, although the Germans knew Patton's army would attack, they did not know when and by which routes (Bowyer 1980: 56).

Modern cases of dazzle include US operations against Libya in 1986. US Navy SEALs secretly landed in Libya from submarines in "deceit and cover" operations to persuade Qaddafi's government that someone was about to attack Libya. The SEALs left Kleenex-brand tissue, matchboxes and pre-smoked Israeli and Soviet cigarettes where Libyan patrols were likely to find them. The operations depended on dazzle. They sought to overload the Libyans with extraneous information so they would not know what would happen next. Would the US attack? Would Israel attack? What were Soviet forces doing secretly in Libya? (Reuters 1987: A5).

Disinformation

Usually conducted at the strategic level, disinformation is a broad category of deception that often relies on camouflage, decoys, dazzle, mimicry and conditioning for success. Disinformation involves both hiding and showing. It seeks to reinforce an existing belief so that ambiguity decreases, making an enemy certain, but wrong. In 1800, the chief of the Prussian General Staff, Helmuth von Moltke, the Elder, said, "Gentlemen, I notice that there are always three courses open to the enemy, and that he usually takes the fourth" (Bowyer 1980: 214–15). In one sense, Moltke was talking about disinformation. The three options that an adversary believes are open to the deceiver can be reinforced to deceive them, even while the deceiver plans to pursue a fourth option. The deceiver attempts to convince the target to develop a hypothesis or theory of the situation that is valid, concrete and well-supported, yet wrong.

Disinformation techniques are ancient. The Mongols forged letters to discredit adversaries and spread rumors to mask their intentions. The Mongols, however, were not immune to disinformation themselves. In 1293, a Mongol-Chinese fleet arrived off Java to attack a rival kingdom. The navigator was uncertain about their location and stopped at a coastal city to ask directions. The helpful people said the target kingdom was a bit farther down the coast. The Mongols sailed on and destroyed that kingdom, although it turned out that the first place they landed had been their intended target. They had destroyed an enemy of their enemy (Bowyer 1980: 26–31).

In the late 1920s, the Soviets conducted a classic disinformation operation that came to be known as "The Trust." Soviet agents, posing as anti-Soviet rebels, contacted anti-Soviet émigrés in the West. At first the émigrés were wary, but the agents showed that they were real anti-Soviets by organizing jail breaks, blowing up police stations and feeding vital information to Western intelligence services. The anti-Soviet operations were well documented and the Soviets were soon able to convince the Russian émigrés and Western intelligence services to provide information on real anti-Soviet individuals and groups in the Soviet Union, ostensibly so that they could cooperate with them. A symbiotic relationship was formed with the Western intelligence services supporting groups they thought were anti-Soviet, while the Soviets gained information about real anti-Soviets in the Soviet Union. The operation lasted eight years and was a major success: for the Soviets. They captured and shot many domestic opponents. The jail breaks, attacks on police stations and information fed to the West had all been carefully staged to gain the West's trust and gather information on anti-Soviet groups in the Soviet Union (Epstein 1989: 22–30).

The Soviets, however, were also the victims of disinformation. In 1937, the Soviet intelligence service asked an agent they had in Germany for information about Soviet generals plotting to overthrow Stalin. Unfortunately for the Soviets, the agent was working for the Germans. The Gestapo took documents from their files about meetings between Soviet and German officers in the 1920s, when the two states were secretly collaborating on various military programs, and changed the dates. They then fed the information through the double agent to the Soviets. Stalin received information that appeared completely realistic, since it was, except that the dates had been shifted forward a decade. It took the Gestapo only four days to make the changes (Epstein 1989: 140–2). The 1937–8 purge that resulted from the supposed military conspiracy against Stalin cost the lives of about 20,000 Soviet officers, more than one-third of those on active duty, and crippled the Red Army. Then Hitler offered Stalin a non-aggression pact in 1939, just when the Soviet Union was at its weakest. Stalin accepted (Andrew and Gordievsky 1990: 139; Epstein 1989: 143). The Nazi disinformation operation was symbiotic to a certain extent in that the Nazis weakened the Soviet Union even as Stalin gained intelligence that allowed him to justify emasculating an internal rival: the military.

Hitler, in fact, was a master of disinformation. In 1933, he sought to project a false image of strength, especially of air power, to deter Britain and France as Germany rearmed in violation of the Versailles Treaties. The weak often turn to deception and Hitler's operation was actually based on Operation Kama, which previous German governments had conducted as they had started to rearm after 1918. In the 1930s, there was already a widespread fear of air attack among British and French leaders, which Hitler reinforced. When British officials toured German airfields, the Germans moved aircraft from field to field so the British would count the planes multiple times in a form of mimicry, thus vastly inflating perceived German air strength. Other ruses were also employed. In 1938, the French Air Chief of Staff was taken up in a German courier plane that had been modified to fly without stalling at the extremely slow speed of 35 mph. A test plane with markings like the new Heinkel fighter then shot past at what appeared to be an incredible speed. The French officer thought that the plane was extremely fast because he was not aware his own plane was going extremely slowly. Given its markings, he also assumed that the Heinkel was a production model, when in fact it was a limited edition aircraft made with rare alloys that could never be put into production. The Germans, in fact, often flew souped up, unique planes at air shows marked as if they were in general production to make foreigners believe that German aircraft were far more advanced than what they were actually producing.

The disinformation worked because seeing was believing but, as Don Quixote said, "Facts are the great enemy of truth" (Epstein 1989: 135). The facts did not represent what they purported to represent, and Germany rearmed in peace. Churchill later said that fear of air raids hurt British policy. In 1936, the British and French did not react militarily to the German remilitarization of the Rhineland mainly due to the fear of air attack, when in fact the German army and air force were weak. Many Luftwaffe planes did not even have any guns. Hitler knew that numbers, not guns mattered. Highlighting the symbiotic nature of deception, the RAF, in part seeking to increase its own budget and build more planes, supported inflated estimates of Luftwaffe strength. The same threat worked during crises that allowed Hitler to take over Austria and the Sudetenland in 1938 without firing a shot (Epstein 1989: 139).

Even after the war started, Nazi disinformation continued. After Poland fell, Hitler needed time to finish rearming. So German intelligence crafted a disinformation story: German generals opposed to Hitler needed a few months of peace to overthrow the dictator. If the British and French attacked, the generals would have to defend the Fatherland and support Hitler. One conduit of the disinformation about the bogus generals' plot was Captain Sigismund Best, who had created a fictional network of spies in Holland and Germany to make money selling information to the British, *á la* Graham Greene's *Our Man in Havana*. Fearing bombing and a repeat of the horrific losses of the First World War, the British grasped at any chance to avoid a total war and played their part in the symbiotic relationship: they decided to give the plotters time. As a result, Hitler got time to rearm and to prepare to attack the West. The plot ended when the Gestapo kidnapped Best and a colleague in Venlo, Holland in November 1939. Even then, the British belief in the validity of the disinformation story was so strong that they still thought German generals might overthrow Hitler (Epstein 1989: 144–8).

The Germans continued to use disinformation throughout the war. In 1941, as they planned Operation Barbarossa, the Germans knew that the massive buildup of more than three million men and thousands of tanks and planes on the Soviet border could not be hidden. Therefore, they devised a disinformation story within a deception operation. The disinformation story for the Soviets was that the German buildup in the east was really a deception operation to deceive the British preparatory to launching Operation Sea Lion, the invasion of England. Unbeknownst to the Soviets, Sea Lion had already been cancelled (Epstein 1989: 122). The disinformation worked because it reinforced Stalin's belief that Hitler would not attack, and Barbarossa achieved complete surprise.

The Germans also used disinformation to conceal the movement of two panzer armies into the Ardennes before they launched a surprise attack on 16 December 1944, which led to the Battle of the Bulge. It was equivalent to moving secretly the entire population, with all the cars and trucks, of Cincinnati, Ohio or Liverpool, England. The operation's name, *Wacht am Rhein* (Watch on the Rhine) was designed to give the impression that it was a defensive operation, reinforcing the Allies' belief that the Germans were only capable of defensive operations by late 1944. All communications were sent by landline, since by then the Germans suspected that the British had broken their codes and were concluding that "all wireless is treason" (Brown 1975: 808). Straw on the roads muffled the sound of military vehicles, which only moved at night. Charcoal was used in field kitchens so smoky wood cooking fires would not betray the presence of large formations in the area. To deter deserters who might tell of the impending attack, SS leader Heinrich Himmler issued an edict that the family of any deserter would be sent to a concentration camp or would be shot. The Germans also set up a phantom army of their own, the Twenty-fifth, north of the Ardennes to confuse the Allies. In Operation Heinrich, a few panzers and a handful of troops of the notional Twenty-fifth Army staged a demonstration by noisily maneuvering near Achen, 40 miles north of where the actual attack would fall (Breuer 1988: 276–9). The disinformation worked and the attack surprised the Western Allies, nearly rupturing their line completely.

The British also used disinformation during the Second World War. Weak, alone and facing a powerful German threat, the British turned to disinformation early in the war to deceive the Germans about their island's strength. In October 1940, Franco's Spanish government asked the British Foreign Office if a Falangist associated with Spanish youth movements could come study the British Boy Scouts. The British agreed, knowing that everything the man saw and heard would reach Berlin soon after he returned to "neutral" Spain. British secret service men, posing as Boy Scout officials, met the Spaniard at the airport and escorted him to the posh Athenaeum Court Hotel. In all of London there were only about four heavy anti-aircraft artillery (AAA) batteries, since most of Britain's small number of anti-aircraft guns were guarding the coast or the approaches to the capital. One of the batteries, however, was moved into a park across the street from the hotel. The puzzled gun crew was ordered to fire continuously through every air raid, regardless of the vicinity of any targets. The racket was terrific and the sleepless Spaniard must have thought there was an AAA gun on every London street corner.

Next the Spaniard was taken to Windsor Castle to inspect a Boy Scout troop. The English offhandedly mentioned that the regiment of battle-hardened troops and large number of tanks guarding the castle were just a small ceremonial guard for the Royal Family. In truth, the regiment was one of the only full-strength units in the British Isles at the time, and the tanks were about half the British Army's entire tank strength. The Spaniard was then shown a harbor, packed with RN ships. His escorts casually mentioned that this was a standard force used to protect most harbors. After the Spaniard left, however, the warships rushed back to their regular assignments all along the south coast. A double agent later intercepted a report to Berlin about the Spaniard's visit, which portrayed the British as living in an armed camp, when, in fact, their island home was almost undefended (Breuer 1988: 73–4).

Later in the war, the British turned their disinformation skills to the goal of masking the timing, location and scale of invasions. One of the more macabre disinformation operations occurred in 1943. After having taken North Africa, the Allies were planning their next invasion and, as Churchill said, "Anybody but a damned fool would *know* that it [the target] was Sicily" (Brown 1975: 279). As Montagu (1969) explained, Operation Mincemeat involved the British releasing a corpse from a submarine so that it would wash ashore in neutral Spain. The corpse, outfitted to mimic a fictitious Major Martin of the Royal Marines, carried papers that suggested that the Allies planned to invade Corsica, Sardinia or Greece. (By giving three choices, the plan was reminiscent of von Molkte's three options, with the enemy choosing the fourth – Sicily.) The Germans received the information from Spanish friends and moved units to Corsica, Sardinia and Greece. These movements left Sicily less well defended when the Allies invaded the island a few months later.

Later in the war, the British also used disinformation to protect their island. During the V1 "Buzz bomb" attacks on London in late June 1944, more than 2,000 anti-aircraft guns were moved to the south coast and one-fifth of the Allies' air power was diverted to hunt V1s. Disinformation, however, was also used at the suggestion of one of Britain's top scientists, Dr R.V. Jones. The Germans were asking agents in England, controlled by the British XX Committee, to report the fall of V1s. If the agents told the truth, it would help the Germans, who were using Tower Bridge as an aiming point. If the agents lied, however, aerial reconnaissance would reveal that they were lying and, therefore, that the British controlled the agents. Furthermore, since the Germans kept track of when they fired each missile, they knew when each impact should be, although they did not know where the missiles impacted. Jones solution was brilliant, if simple. He proposed that the agents report on the missiles that had fallen north of London using the impact times of missiles that had actually fallen south of the capital. Therefore, the Germans would conclude that the flying bombs were falling long and shorten the range by decreasing their flight times. The missiles would then fall in open country south and east of London (Brown 1975: 723–4; Latimer 2001: 201–2). The disinformation was sent via the double agents and the V1s started to fall shorter, saving the crucial docks along the Thames and government offices in Westminster. The poorer areas of East London, however, paid a price for the disinformation campaign.

Japan also used disinformation. In the 1930s, Japan deceived the West by staging air shows in which their pilots made mistakes on purpose and her air force officers complained to Westerners about the difficulty Japanese pilots were having flying solo. These comments reinforced the common Western belief that the Japanese were far from competent pilots, thus making the West less likely to take the increasing threat from Japan seriously (Epstein 1989: 123–4).

After the Second World War, the Western Allies, for the most part, dropped disinformation from their arsenals. The Soviets did not. Soviet maps, for example, were falsified by deleting major streets and omitting scale. Street maps of Moscow also deleted many important buildings, including the KGB building on Dzerzhinski Square (Gerwehr and Glenn 2000: 21–2). The KGB's First Directorate, the disinformation section, produced bogus memoirs and fake letters. Some were silent forgeries that were only shown in confidence to influential Third World figures to alert them to allegedly hostile US military and intelligence operations. Other stories were disseminated via the media, often originated by KGB officers working in the Novosti Press Agency. For example, in 1968 the KGB forged a letter from Gordon Goldstein of the US Office of Naval Research supposedly revealing US bacteriological warfare in Vietnam and Thailand. It appeared as a story in the Bombay *Free Press Journal* and then in the 7 March 1968 London *Times*. It was later found that the letterhead and Goldstein's signature had been copied from a conference invitation sent to scientists. Forged documents also were used to support charges that the United States used germ warfare during the Korean War. Even though they were Soviet disinformation, stories about US germ warfare in Korea and Vietnam enjoyed long lives (Andrew and Gordievsky 1990: 503–4). In September 1983, the KGB started a story that the Korean Air Lines (KAL 007) airliner that the Soviets shot down over Kamchatcka was on a US spy mission. There was no validity to the story, but it gained widespread attention (Andrew and Gordievsky 1990: 593–8). The KGB also forged a 1987 letter from CIA director William Casey that discussed plans to destabilize the Indian government, a 1988 National Security Council document containing instructions from President Reagan to destabilize Panama, and a 1989 letter from South African Foreign Minister "Pik" Botha to the US State Department referring to a secret agreement for military, intelligence and economic cooperation with the United States (Andrew and Gordievsky 1990: 630).

One of the KGB's most successful disinformation operations started in 1983 in an article published in the pro-Soviet Indian newspaper, *Patriot*. The article alleged that the US Department of Defense had created HIV, which causes AIDS, during genetic engineering experiments at Fort Detrick, Maryland. Initially the story had little effect, but it was revived by the Russian *Literaturnaya Gazeta* in October 1985. The story was supported by a report from a retired East German, Russian-born biophysicist, Professor Jacob Segal, which sought to demonstrate that the virus was synthesized at Fort Detrick from two natural viruses. The report has since been discredited, but thus supported by quasi-scientific jargon, the story swept through the Third World and even into the West. In October 1986, the conservative British *Sunday Express* made an interview with Segal the basis for a front-page story, while in the first six months of 1987 the story received major coverage in more than 40 Third World countries. The world scientific community, including the Soviet Union's leading AIDS expert, Viktor M. Zhdanov, discredited the story and in August 1987, fearing embarrassment if their disinformation operation was revealed, the Soviets told US officials that they had officially disowned the story. Soviet press coverage of the story came to a halt, but the story is still circulating on the Internet, such as at http://www.apfn.org/apfn/aids.htm (accessed 24 March 2006), in the Third World and sometimes in the West. For example, Segal and Fort Detrick were featured in a documentary about AIDS made by a West German firm in January 1990 for Britain's Channel Four and Deutsche Rundfunk WDR, Cologne (Andrew and Gordievsky 1990: 630–1).

The KGB enjoyed similar success with a story about Americans dismembering Latin American children to use their body parts for organ transplants. In 1988 the story was picked up by a Brussels-based, Soviet front organization, the International Association of Democratic Lawyers (IADL), who publicized the story extensively in more than 50 countries. In September 1988, a French Communist member of the European Parliament (EP), Danielle de March, even proposed a motion condemning alleged trafficking in "baby parts" and cited an IADL report to support her charges. The motion passed by a show of hands in a poorly attended EP session. A Greek newspaper reported that human hearts were for sale in the United States for between $100,000 and $1 million each. Not limiting themselves to HIV and organ transplants, the KGB also planted stories in the 1980s about US development of an "ethnic weapon" that would kill only non-whites (Andrew and Gordievsky 1990: 631–2).

The Soviets also used disinformation to disguise the accuracy of their missiles. In the early 1960s, the Soviets broadcast bogus telemetry to make their SS-7 missiles appear less accurate. The goal was to reinforce the US belief that American missiles were safe from attack. The Soviets also allowed an agent to be caught attempting to purchase accelerometers in the West, which are crucial to missile accuracy. This was meant to suggest that the Soviets were working on improving the accuracy of their, presumably, inaccurate missiles. On their test ranges the Soviets left telephone poles in place as targets so that US satellites would photograph impact craters far from their telephone-pole targets. The Soviets, however, moved the poles and filled in or created craters as needed to make their missiles appear far less accurate than they actually were. Soviet moles in the US Joint Chiefs of Staff served as the feedback mechanism for the operation to ensure that the Americans bought the story. By reinforcing the American belief that US technology was superior to Soviet technology, the disinformation campaign worked (Epstein 1989: 165–76).

The United States also has been the target of more recent disinformation operations. The Egyptian media accused the United States of dropping poisoned food packets in Afghanistan in 2003–4, and of spreading AIDS in Africa (Boot 2005: B13). In a case with far greater effects, the Bush administration's rationale for intervening in Iraq in 2003 was largely based on the belief, since disproven, that Iraq was close to developing weapons of mass destruction (WMD). It was later revealed that a Bush claim that Iraq had sought uranium from Niger was based on forged documents, while reports of an Iraqi fleet of mobile laboratories designed to manufacture biological weapons was based on a highly unreliable, German-controlled source, code-named Curveball, who even the Germans said was "crazy." It also turned out that at least two exiled Iraqis, intent on persuading the United States to invade Iraq, had provided disinformation to the United States. The exiles suggested links between Al Qaeda and Iraq and, more importantly, that Iraq was close to developing WMD. The conclusion the exiles and Curveball helped reinforce before the invasion, that Iraq was about to become a nuclear, chemical and biological threat to the United States, was incorrect (Drogin and Miller 2005: A1, A19; Miller and Drogin 2005: A1, A13). Members of the Bush administration were far from the only ones to have their beliefs reinforced by disinformation about Iraq. Many news organizations, including the *New York Times*, were taken in by sources providing inaccurate information (Bennett and Weiskopf 2004: 36–7; Drogin 2004: A1, A7; Reuters 2004: Online; Rutten 2004b: E1, E14; The [*New York Times*] Editors 2004: A10). It is unclear who was behind the disinformation, but it is clear that many parties were attempting to deceive the United States.

Other states have also employed disinformation. At the start of the 1973 October War, for example, the Egyptians conducted 150 economic, political and military deceptions in a massive disinformation campaign to make the Israelis certain, but wrong in their belief that the Arabs would not attack. Some of the Egyptian staff who devised the disinformation campaign had served on Montgomery's staff at El Alamein, and they put their experience in deception to good use. Before the attack, the Egyptians demobilized 20,000 troops to make it appear as if an attack was not imminent, staged numerous canal-crossing exercises to get the Israelis used to seeing preparations for a crossing, and used construction activities to mask canal-crossing equipment near the canal. They spread rumors that made an attack seem unlikely and continued to build defenses along the canal to convince the Israelis that Egypt was not planning to attack (Gerwehr and Glenn 2003: 21–2, 43–4; Latimer 2001: 85–8; Raviv and Melman 1990: 206). The Egyptians also sparked three separate crises before the actual attack by moving forces up to the canal and increasing their anti-Israeli propaganda, which forced the Israelis to go on alert in May, August and late September. In October the Israelis, therefore, were far less likely to go on alert when they noticed preparations for an attack that appeared similar to the preparations made during the three earlier crises, which had not led to war. There was ample information that pointed to an impending attack, yet Israeli decision-makers did not go against their belief that an Arab attack was highly unlikely, and the Arabs, by reinforcing the Israelis' belief, achieved surprise. Israel ordered only a partial mobilization on 5 October, the day before the Arabs attacked (Westwood 1984: 136).

Conditioning

Conditioning involves creating a pattern of behavior so that the target develops an expectation that later can be exploited. The Egyptians' use of three crises before their actual attack in 1973 was a case of conditioning in the sense of "cry-wolf" tactics. The Egyptians desensitized the Israelis to a pattern of Egyptian behavior (moving troops and equipment up to the canal) and then used that desensitization to surprise the Israelis when they moved troops and equipment up to the canal and attacked.

Conditioning, however, is far older than 1973. In 212 BC, Hannibal took the city of Tarentum from the Romans by using conditioning. Hannibal had a dissident Greek resident of Tarentum, Cononeus, create a nightly routine of leaving the city to go hunting with a large party. The Roman guards grew used to the hunt leaving each night and in the symbiotic aspect of the relationship, liked the meat the hunters brought back for the garrison. Then one night, after the Romans were well conditioned to the behavior, Hannibal disguised some of his best warriors as hunters. They returned as Cononeus' party, overpowered the guards and opened the city's gates to Hannibal's army, which conquered the city with few casualties (Gerwehr and Glenn 2000: 22).

As in the case of the Egyptians in 1973, disinformation often relies on conditioning. A spy, for example, may provide valid information for months or even years, conditioning their handler to believe the information, before the spy provides the single key piece of disinformation. The British conditioned the Germans in this way when they ran double agents during Operation Bodyguard to deceive the Germans about the Normandy landings. The British fed largely valid information to the Germans for years before providing disinformation in the spring of 1944 about the landings.

Change rules/tactics

Julius Caesar wrote, "The most powerful weapon of war is the unexpected" (Brown 1975: 415). The quote suggests another deceptive stratagem that involves misleading an adversary by doing something unexpected, such as using a new weapon, sensor or tactic, or doing what appears to be impossible.

A new weapon or sensor can, by its very newness, deceive an adversary. Gunpowder, mines, poison gas, submarines, tanks, radar, sonar, stealth technology and a multitude of other weapons and sensors were deceptive in that when they were introduced, the other side was surprised. It can be easy to deceive an adversary because new weapons can be hard to see even when they are in plain view. British photo-interpreters, for example, failed to recognize German V-2 rockets in reconnaissance photographs in part because the British expected the rockets to be much larger (Keegan 2003: 22).

A new invention does not have to be turned into a weapon to be used to deceive. For the landings at Normandy, the Allies devised floating, mobile harbors, called Mulberries, to tow across the Channel. They also developed a Pipe-Line Under The Ocean (PLUTO) to supply the Anglo-Canadian-American armies on the beachheads with vital fuel. The new inventions meant that the Allies did not have to land at or near a port, which greatly expanded the number of possible landing beaches. The Mulberries and PLUTO thus helped deceive the Germans about where the Allies would land.

A deceptive innovation can also take the form of a new tactic. The Greeks, known for fighting their adversaries face-to-face on the battlefield, changed their tactics when they used the Trojan horse. Not known for guile, the Greek's use of deception, in and of itself, deceived the defenders of Troy. Similarly, when Goliath strode forth to do battle in 1013 BC, he expected to face another warrior in single combat with helmet, shield and sword. David's use of a standoff weapon, a slingshot, was a change in tactics. Therefore, the use of the slingshot was deceptive and a surprise. It also won the battle for David. German U-boat wolf pack tactics during the Second World War are an example of using established weapons in a new way, as was the German use of dive bombers, tanks and mechanized troops coordinated by radio in blitzkrieg operations in 1939–40. While the French and British parceled out their tanks in small numbers to almost all of their divisions, the Germans concentrated their armor in panzer divisions with several hundred tanks and armored vehicles in each. Radio was used to coordinate the tanks' movements and to call in close air support. It was an innovation that was deceptive in that it was unexpected. It was also extremely effective.

The British use of radar in 1940 was also deceptive in that, instead of just using radar to detect German aircraft as they took off from France, the British had created an entire system to control their fighter aircraft. Radar meant that the British knew how many German aircraft were attacking, where and when. Therefore, the British could send an appropriate number of fighters to intercept. There was no need for the RAF to maintain aircraft in the air awaiting attack with the commensurate wear and tear on pilots and aircraft. The radar, by itself, would have helped the RAF somewhat, but the intricate control system was crucial to the British victory in the Battle of Britain. The Germans, deceived about the true importance of the radar system, did not realize its battle-winning significance until it was too late.

Doing what appears to be impossible is also a deceptive stratagem. In 1759, for example, Wolfe sent his British troops up a goat trail to climb the "unscalable" Heights of Abraham to attack the French forces under the Marquis de Montcalm in Quebec City. Montcalm was shocked to see the British forces suddenly assembling on the Plains of Abraham before the walled city. When the French came out to fight, they were defeated. Wolfe had won by maneuvering beyond Montcalm's conception of the battlefield and doing the "impossible" (Bowyer 1980: 37). Similarly, in 1950, MacArthur went beyond the North Korean's conception of the battlefield by landing at Inchon far behind the battle lines around Pusan on the southeast tip of Korea.

Going beyond the conception of the battlefield can also relate to beliefs about technology. For example, the German belief in the Second World War that the Enigma code could not be broken, allowed the Poles, French and British, in a tag-team-like effort, to break the code and yet continue to deceive the Germans by reinforcing their belief that their codes were secure. Of course, any case of code breaking from the American breaking of Japanese codes before the battle of Midway to the Axis breaking of Allied codes, as long as it is kept secret, allows the code breakers to deceive their adversaries by reinforcing the adversary's belief that their codes are secure or even unbreakable.

8

THE BEST DECEIVERS

The British in the Second World War

Early in the Second World War, the British weak, outnumbered, alone and with no wish to repeat the bloodbaths of the Somme, Passchendaele and Ypres, took deception operations to new heights of planning and implementation. General Sir Archibald Wavell, C-in-C Middle East, was a key figure in the development of the London Controlling Section (LCS), which ran British deception operations globally during the war. The LCS would become so crucial to victory, even more so than the atomic bomb, that documents related to it were kept secret until 1975, when the US National Archives released a single document mentioning such operations (Brown 1975: 47).

In 1940, Wavell wrote a brilliant memorandum in which he outlined the principles of deception and recommended the creation of a governing authority to run global deception operations. He discussed visual, aural and even nasal ruses, as well as "signal deception" for wireless communications, which was, then, just coming into existence. For deception operations to succeed, he stressed, it was crucial to know the enemy commander's mind: "It will be far more valuable to know that an opposing general is thoroughly excitable than it is to know that he graduated at some staff college in such-and-such a year" (Brown 1975: 47).

Wavell was not just a theoretician and he used deception masterfully when the Italians declared war on 10 June 1940. With 36,000 men and an incomplete tank division, Wavell faced 200,000 Italians in Libya and another 110,000 in the Red Sea states. Wavell summoned Brigadier Clarke, then 41, to help create "A" Force. Clarke has been called the father of modern deception operations, setting out most of the principles of deception that would be used for the rest of the war and ever since. Even though it was never large, at its height having just 41 officers, 76 NCOs and three units of company strength, "A" Force had a huge effect (Latimer 2001: 119–28).

When the Italians attacked on 13 September 1940, "A" Force used inflatable rubber tanks, field guns and 2-ton trucks to mimic a powerful phantom force just south of the main Italian attack. Arabs with camels and horses dragged harrow-like devices behind the rubber decoys to create great dust clouds, while British flak guns kept Italian planes at high altitude, so they could not get a close look at Clarke's rubber army. In response, Marshal Rodolfo Graziani concluded that there was a powerful British force on his right flank. Fearing a counterattack, the Italians halted their attack and dug in. Wavell then received real reinforcements and counterattacked on 9 December, advancing 650 miles into Libya. By 7 February 1941 he had captured 130,000 prisoners, 400 tanks and 1,290 guns at a cost of 135 dead, 310 wounded and 52 missing. The victory destroyed Italian hopes for an African empire and shattered their morale, arguably for the rest of the war (Bowyer 1980: 84–5; Brown 1975: 49–50).

In a crucial development, Clarke devised a Notional Order of Battle, which was a fictitious list of forces the Allies had in a given theater, usually far more than were actually present. "A" Force used real units that were not yet operational as the basis for many of their notional units, thereby following the deception rule to use as much real information as possible for deception operations. The goal was to have the enemy believe the Notional Order of Battle, supported by bogus air fields, supply dumps and radio traffic, and deploy their forces inefficiently. The Notional Order of Battle was extremely effective. By the late spring of 1944 the Germans believed that the British and Americans had 14 divisions in Libya and Egypt alone, when in fact the Allies only had three divisions in those countries. The phantom Allied force was seen as a threat to the Balkans, where the Germans kept 22 divisions awaiting an Anglo-American attack that would never come (Holt 1994: 49–50, 54; Latimer 2001: 120). Clarke was so good at creating fake armies that at one point, when the Italian campaign was going poorly for lack of infantry, Eisenhower was not pleased to learn that the Ninth and Tenth British Armies, which were in reserve, did not exist (Brown 1975: 116–17).

One of "A" Force's greatest successes was at the Second Battle of El Alamein in October–November 1942. General Irwin Rommel knew that the only place the British could attack was in the north. The impassable Qattara Depression cut off movement 40 miles south of the Mediterranean. To make matters worse, the desert near El Alamein was a plain of hard sand, so any movements were easily visible from the German lines. General Sir Francis Wilfred de Guingand, Montgomery's Chief of Staff, told Clarke;

> Well there it is. You must conceal 150,000 men with a thousand guns and a thousand tanks [as well as thousands of vehicles and tens of thousands of tons of supplies] on a plain as flat and as hard as a billiard table, and the Germans must not know anything about it, although they will be watching every movement, listening for every noise, charting every track. Every bloody wog will be watching you and telling the Germans what you are doing for the price of a packet of tea. You can't do it, of course, but you've bloody well got to!
>
> (Brown 1975: 117–18)

Clarke turned to Lieutenant Colonel Geoffrey Barkas, a film set designer, and Major Jasper Maskelyne, a conjurer. They decided to do what Malcolm had done at Birnam Wood; move forward under camouflage so gradually that the Germans would fail to perceive the movement. It would "become one of history's most remarkable conjuring tricks" (Brown 1975: 118). They sought to hide the British intention to attack in the north by minimizing the scale of the apparent buildup and making it appear that they were not yet ready to attack. To start with, the British hid 2,000 tons of petrol in the north within 5 miles of the front in existing slit trenches. By lining the gas cans in the trenches, aerial reconnaissance photographs showed no difference in the shadows in the trenches than when they were empty. Then 4,000 tons of stores were brought in over three nights and stacked in a way that they could be covered with nets and appear to be benign 10-ton trucks. Hiding 1,000 guns and their prime movers (tractors) was especially difficult, since they had a distinctive shape. "A" Force, however, found they could fit a prime mover and its gun under a dummy of a 3-ton truck. Three thousand guns, limbers and movers were hidden as 1,200 trucks, which was not an uncommon sight in the desert. When the guns moved forward on the eve of the attack, they moved at night and were replaced by 1,200 more dummy trucks so the Germans would not know that they had moved at all.

The armored divisions were harder to conceal. In a conditioning process, the divisions' soft-skinned vehicles were moved in and out of the area of the attack several times over three weeks so the Germans would grow accustomed to their movements. The 720 tanks, self-propelled guns and armored cars were the most difficult problem. The armor was massed at three points, Murrayfield (North), Murrayfield (South) and Melting Pot, where several desert tracks converged. These tracks tended toward the south, where the British wanted the Germans to think an attack was coming, but nearby was a second series of tracks that led north to where the British would attack. The Germans noticed the concentration of armor but assumed the tanks were headed south. The night of the attack, the armor headed south, then doubled back to their staging areas in the north. Once in attack position, each tank slid under a "sunshield" that resembled a 10-ton truck and had been in position for a week to lull the enemy into a false sense of normalcy. To hide the fact that the armor had moved, dummy armor replaced it at the three staging areas and palm fronds were used to obliterate the tanks' tracks. By dawn, the three staging areas looked exactly as they had the day before; like a tank force waiting to attack in the south (Brown 1975: 118–19).

To support the supposed attack in the south, "A" Force even built a dummy pipeline toward the southern front. A trench was built in standard fashion, five miles at a time, and a dummy pipeline made of old four-gallon petrol cans was placed alongside the trench. Then, at nightfall, the trench was filled in and the cans were moved forward, conforming exactly to the rate at which an actual pipeline would be built. The plan was devised so as to appear that the pipeline would not be completed until 48 hours after the British had launched their attack in the north. Three dummy pumping stations and supply dumps were also constructed in the south to support the supposed attack. There was also a double bluff: telephone-pole gun positions in the south were camouflaged, but the nets allowed to rot. The Germans would see that they were fake, yet the night before the attack in the north, real troops would arrive with a few guns. In the morning the "fake" guns would speak, hopefully convincing the Germans that the attack in the south was real. Ultra intercepts of German radio messages confirmed to the British that the Germans had been deceived and the attack achieved surprise (Brown 1975: 119–23).

Realizing the importance of deception, in April 1941 a group of British intellectuals began meeting in a conference room in Churchill's underground headquarters beneath Storey's Gate, London. The group was one of the most closely guarded secrets of the war and was known by the innocuous title of the London Controlling Section (LCS). The LCS developed stratagems and coordinated the activities of the myriad British deception and cloak-and-dagger agencies.

> LCS was the first bureaucracy created in Great Britain's history whose sole function was deceit. In implementing stratagems to deceive the Germans, LCS would meld techniques used for centuries with the special means of modern technology and psychological and economic warfare, using any and all arms of the government and the military. LCS methods would not only be secret and intricate, but ruthless. Under its chairman, Colonel Oliver Stanley, a prominent Tory minister, the LCS would see that each British war-planning staff had a deception section linked directly to LCS. As the result of this network, the sum of all intelligence data could be utilized by LCS to create stratagems that would have enormous impact on current and future military operations.
>
> (Breuer 1988: 95)

By the end of the war, the British had more than 100,000 men and women employed in intelligence and propaganda work (Brown 1975: 34), and their greatest operation and greatest success was Operation Bodyguard.

For the invasion of Normandy, Operation Bodyguard sought to mask the concentration and movement of an entire army comprising eight sea- and airborne divisions numbering about 200,000 men. Three reserve divisions of 90,000 men and some 19,000 vehicles would follow the assault elements. The force would land from 5,000 assault ships and 1,300 merchantmen, protected by more than 1,200 warships and more than 10,000 aircraft. More than four million men and women were involved in the D-Day operation. The invasion force was also only the tip of a force that would total some six million British, Canadian, American and allied men and women (Brown 1975: 413).

"Hitler had the strength to repulse Neptune [an early name for Overlord], unless the Allies could disperse his forces, take him by surprise on D-Day, and confuse and delay his reaction to the assault" (Brown 1975: 413). The topography of the coast, the tides, the phases of the moon and the weather, however, greatly narrowed the possible locations and times of the invasion. And, once the invasion began, it seemed impossible that Hitler would not be certain that Normandy was THE invasion, yet that was, in part, Bodyguard's goal.

If the deception plan failed, the Allies had cause to worry based on recent experience. In January 1944, the Allies landed at Anzio, Italy, and in a month of fighting suffered 19,000 casualties. The Luftwaffe pounded the beachhead even as panzers counterattacked with German infantry, especially paratroops, which fought with skill and tenacity. The beachhead was hemmed in and the capture of Rome long delayed (Brown 1975: 422).

Churchill was largely responsible for pushing the deception operation for Normandy, first called Operation Jael (after a woman in the Bible who invited an enemy into her tent and then killed him by driving a tent peg through his head as he slept) and, later, Bodyguard (Brown 1975: 804). While in the political wilderness between the wars, Churchill wrote;

> Battles are won by slaughter and manoeuvre. The greater the general, the more he contributes in manoeuvre, the less he demands in slaughter... Nearly all the battles which are regarded as the masterpieces of the military art... have been battles of manoeuvre in which very often the enemy has found himself defeated by some novel expedient or device, some queer, swift, unexpected thrust or stratagem. In such battles the losses of the victors have been small. There is required for the composition of a great commander not only massive common sense and reasoning power, not only imagination, but also an element of legerdemain, an original and sinister touch, which leaves the enemy puzzled as well as beaten.... There are many kinds of manoeuvres in war, some only of which take place upon the battlefield.... There are manoeuvres in time, in diplomacy, in mechanics, in psychology; all of which are removed from the battlefield, but react often decisively upon it, and the object of all is to find easier ways, other than sheer slaughter, of achieving the main purpose.
>
> (Brown 1975: 5)

Largely based on Churchill's support, Bodyguard was essentially a British operation. In part this may reflect that Britain was weaker militarily and, therefore, more inclined to rely on guile rather than force, while the United States was militarily more powerful and more inclined to rely on force rather than guile. The British also had more experience using deception in their empire than did the Americans domestically.

The Allies had seven variables they could manipulate to deceive the Germans about the Normandy invasion: actors, means, intention, consequence, location, time and strength (Bowyer 1982: 86). The Allies convinced the Germans that they had to worry about possible Turkish entry into the war, for example, thus deceiving the Germans about the possible actors who would be involved. Casting Patton and Thorne as possible commanders of the invasion also further deceived the Germans about the cast of characters. The British also were deceptive about the means by which they would invade. They concealed the construction and use of the Mulberry Harbors and PLUTO, as well as new amphibious trucks, flail tanks to detonate mines, and flame-throwing and bridge-carrying tanks.

A range of plans deceived the Germans about Allied intentions and consequences. Plans to land in the Balkans reinforced a German belief that the British and Americans might seek to deprive the Soviets of gains in the Balkans. Plans to land in southern France also suggested that the Allies planned as a consequence to link up with their forces in Italy and end the war by invading Germany from the south. Furthermore, propaganda and agents fed the Germans disinformation on plans to rely on bombing to win the war without any invasion.

Just as Sherman had done during his drive on Atlanta, the Allies devised their plans based on the principle of offering the adversary a choice (Bowyer 1982: 86–8). For location, the Allies tried to convince the German high command that landings might occur in Brittany, the Pas de Calais, Holland, the Balkans or Norway. As to the timing of the invasion, the Allies provided information that suggested that the invasion would be later than 6 June. This information was based on rehearsal schedules and on a bogus Allied belief that they would need 50 divisions to invade France. Such forces would not be available until the summer of 1944. The British also fed the Germans information that the Soviets would not start their summer offensive until late June, so Hitler would be uncertain of which front to reinforce or when to do it. In relation to the strength of the invasion, it was of paramount importance that "A" force had already spent years developing a Notional Order of Battle. The Allies used the notional forces to give the impression that the invasion force was larger than it actually was, in part to suggest that Normandy was not the main attack and that there would be other landings after D-Day (Brown 1975: 433–4).

Bodyguard was an overall strategy consisting of 36 subordinate plans and scores of associated stratagems. Various plans sought to convince the Germans that the Allies would invade elsewhere, including Zeppelin (southeast Europe), Fortitude North (Norway), Fortitude South (Pas de Calais), Ironside (Biscay), Vendetta (Marseilles coast) and Copperhead (southern France). The Soviets were also involved, making feints in Karelia (in Northern Russia near the Gulf of Finland) and against the Black Sea coast of Rumania. Diplomatic and political actions against Hitler's allies in Eastern Europe and the Balkans were also undertaken to divert attention from Normandy.

> In all, the plans were clever, logical and plausible; but on paper they were about as exciting as a large-scale corporate fraud, which, in a sense, they resembled – a skillful juggling of facts by one group of experts to deceive another group.
>
> (Brown 1975: 436)

In fact, the British used only a few select agents to supply the crucial information. If they used too many agents and the Germans figured out that one was a double agent, the entire deception story would be discredited. Information was leaked slowly and carefully by operatives called "musicians and choristers," which suggested the need for near-perfect coordination. The British ensured that the Germans worked hard to gather the information, because "intelligence easily obtained is intelligence readily disbelieved" (Brown 1975: 436).

Deception was used in conjunction with many other techniques. Offensive intelligence sought to learn German secrets and provide feedback on how the Allied deception story was being received. Ultra was crucial in this area, as was the *Schwarze kapelle* (Black Orchestra), which was a small group of anti-Nazi German officers and men that supplied intelligence to the Allies. Counterintelligence and security measures sought to deny Hitler knowledge of Allied intentions. The XX Committee ran German double agents in Britain, which were crucial to keeping the invasion secret by closely controlling the information Hitler received from agents in Britain. To ensure security, only an extremely small group of men knew about Bodyguard. To ensure coordination, men at lower levels had to follow orders precisely, something some, especially the Americans, sometimes failed to do, several times jeopardizing the entire deception operation (Brown 1975: 276). Special Operations agents sought to locate, encourage and supply guerilla groups across Europe to be ready to support the invasion. Political warfare, including the British Political Warfare Executive (PWE) and the American Office of War Information (OWI), both of which specialized in propaganda, also supported the invasion.

Although Bodyguard was a strategic deception, many tactical measures also were taken to deceive the Germans. During D-Day, for example, the Allies dropped 200 half-sized paratrooper dummies, which the Germans called "dolls." The dolls were accompanied by thousands of rifle and machinegun simulators, pintail bombs that shot Very lights (colored flares shot from a special pistol named after the US inventor, E.W. Very) into the sky, and special forces units that played tapes of the sounds of mortars, soldiers talking and swearing, and rifle fire. It was all designed to give the impression that thousands more paratroopers were descending on Normandy than in fact were (Brown 1975: 647–8).

As Brown (1975) showed, in coordination with these other techniques, deception planners planted thousands of scraps of information, which, when assembled by the Germans, would lead them to form a plausible and acceptable, but incorrect, picture of Allied military intentions. All of the facets of the deception plans with the supporting offensive intelligence, counter intelligence, security, special operations and propaganda rested on each other like a house of cards. Alone, they could do little, but taken together they formed an impressive structure that deceived the Germans and helped make the Normandy invasion a success.

Even with all of the extensive planning, before the landings there were grave doubts about whether Bodyguard would deceive the Germans. In 1943, the US Joint Chiefs of Staff wrote, "We feel that the over-all [*sic*] deception policy in Plan Jael [Bodyguard's early name] is so ambitious as to be the subject of some question as to its general plausibility" (Brown 1975: 437). Surprise and, therefore, deception were crucial to the success of the landings. Without surprise, the German army was still strong enough to destroy an Allied landing and change the course of the war, at the minimum greatly extending the life of the Third Reich. The D-Day deception operations, however, proved to be extremely effective in dispersing German dispositions, achieving surprise and reducing Allied casualties. Furthermore, after the landings, Allied commanders acknowledged Bodyguard's significant contribution to victory at Normandy.

The key to any amphibious landing is to build up forces faster in the beachhead than the defender can bring in to contain and destroy the beachhead. One of Bodyguard's goals was to ensure that the Germans dispersed their forces so they would not have the units near enough to Normandy to attain local superiority and push the invaders back into the sea. As one British official wrote:

> There was always the necessity to do everything possible to induce the enemy to make faulty dispositions of his . . . reserves, to strive if possible to have him at a disadvantage. Damned unsporting, of course, but that is the way in real war.
>
> (Brown 1975: 439)

Most of Bodyguard's sub-plans worked in this respect. For example, Fortitude North, the fictional threat to Norway, caused Hitler to reinforce his garrison in the northern land. By the late spring of 1944, the Germans had 13 army divisions, with 90,000 naval and 60,000 air personnel, 6,000 SS and 12,000 paramilitary troops stationed in Norway (Brown 1975: 472). Plan Zeppelin, which suggested that the Allies would land in the Balkans was also effective. On D-Day, Hitler had 25 divisions in the Balkans, including two powerful panzer divisions. A further 18 infantry, SS and paratroop divisions, as well as seven panzer divisions were bogged down fighting in Italy. The forces Hitler kept in the Balkans and Italy accounted for one-sixth of his forces. Even one division in Normandy on D-Day might have changed the outcome completely, but Bodyguard ensured that Hitler was never certain where the blow would fall around the extended perimeter of his *Festung Europa* (Fortress Europe) (Brown 1975: 404).

In May 1944, German deployments were as follows; Russian Front, 122 infantry, 25 panzer, and 17 other divisions; Italy and Balkans, 37 infantry, 9 panzer, 4 miscellaneous divisions; Western front including the Netherlands, Denmark and Norway, 64 infantry, 12 panzer and 12 miscellaneous divisions; reserves in Germany, 3 infantry, 1 panzer and 4 miscellaneous divisions. In France on D-Day, Hitler only had 50 infantry and parachute divisions, and 10 panzer divisions. In part in response to Bodyguard, the Wehrmacht had dispersed its forces and, more importantly, it had almost no reserves to respond to an invasion. Furthermore, even the units that were in the West were concentrated in the wrong place: Pas de Calais.

More importantly in some ways, and more directly attributable to Bodyguard, was Hitler's belief that there would be secondary landings elsewhere before the main landing at the Pas de Calais. This belief allowed the British to continue to deceive the Germans about the true nature of the Normandy landings – that they were THE main landing – even after the Canadians, British and Americans were well established ashore. The continuation of Bodyguard after the Normandy landing kept at least 20 divisions in the Pas de Calais for months after the invasion. This was crucial to the landing's success, since the Germans committed their forces to Normandy piecemeal and not all at once. As General Omar Bradley later wrote, Hitler was "paralyzed into indecision in Normandy by the conviction that he had more to fear from Calais" (Brown 1975: 804). Furthermore, fear of other possible landings further hindered the German response to the invasion. For example, even after D-Day, Operation Vendetta reinforced the German belief that the real landing would be elsewhere, possibly the south of France. Of 10 German divisions in the south of France, only one was sent to fight in Normandy (Holt 1994: 55–6).

Thanks in part to the deception plan, but also to the weather, the invasion achieved surprise. The weather played a crucial role since the Germans did not believe that the Allies would land in bad weather (Brown 1975: 670). Eisenhower, however, decided to invade between a series of storms that raged across the Atlantic in early June 1944. He had a major advantage over the Germans. Allied weather stations on Greenland and Iceland, and reports from Allied ships in the Atlantic were able to present him with relatively accurate predictions about gaps in the storms. The Germans, who had no stations west of the French coast, had much less, if any such information on the weather fronts coming across the Atlantic. They believed that the weather on 6 June all but ruled out an invasion; a belief only reinforced by Bodyguard's story that the Allies were not ready to invade yet in early June.

The deception operation helped keep casualties relatively low at the British (Gold and Sword), American (Utah and Omaha) and Canadian (Juno) beaches. During the invasion's first day, the British, Canadians and Americans lost 12,000 men killed, wounded and missing. If Bodyguard had failed and surprise had not been achieved, Allied planners had expected 75,000 casualties (Brown 1975: 678). At Dieppe, the Canadians suffered 3,379 casualties, with another 797 Royal Navy and British Commando casualties. This totaled more than a quarter of the casualties suffered by the Allies on D-Day by a Dieppe force that was tiny (about 2 percent) compared with the vast number of troops who invaded Normandy.

In the first seven weeks after D-Day, the Americans suffered 133,326 casualties (34,133 dead), while the British and Canadians suffered 83,825 casualties (16,138 dead). Many more would have died if the Germans had not been deceived. German reinforcements that could have wrecked havoc were nearby, but were not thrown into the battle because of Bodyguard. The Germans still kept their powerful Fifteenth Army in the Pas de Calais area to defend against an attack they thought was coming from the phantom FUSAG, which appeared to be still in England (Breuer 1988: 251–2). Such German reinforcements could have had a telling effect on the outcome of the invasion. When the Allies landed at Salerno on 8 September 1943, the Germans reacted quickly. The US commander, General E.J. Dawley, hesitated, and the Americans were pinned down on the beaches. The operation cost about 15,000 US casualties, far more than expected. It took 23 days for the Allies to take Naples, which they had expected to seize rapidly. Dieppe and Salerno showed how bloody an invasion could be if the defenders reacted rapidly with all the forces at their disposal, something Bodyguard ensured the Germans did not do at Normandy (Brown 1975: 298–9).

Besides dispersing German forces, achieving surprise and keeping Anglo-American-Canadian casualties lower than expected, the Allied commanders also concluded that the deception plan had been successful. Eisenhower signaled General George C. Marshall on 6 July 1944 that Bodyguard had been "remarkably effective" and had "paid enormous dividends." In his 1945 Report to Congress, Eisenhower wrote that the Allied armies on D-Day had "achieved a degree of tactical surprise for which we had hardly dared hope" (Brown 1975: 804). Patton concluded in October 1945, "I feel that deception and cover plans or operations are fully justified and that the employment of cover and deception should…be an accepted and organized procedure for any campaign" (Mullen 2005: Online). As Churchill had said at the Tehran Conference, "a bodyguard of lies" (Brown 1975: 389) had protected the truth and deception had triumphed. Bodyguard had helped to disperse German forces, achieve surprise, delay German reinforcement of the crucial sector, save Allied lives and contributed significantly to the success of the landings on the beaches of Normandy.

9

THE EASIEST MARK

The United States

A number of changes in the media have increased the ease with which a bogus story based on altered images, whether for propaganda or to deceive, can be spread far and wide. The television news, especially CNN, sets the agenda for much of the rest of the media. Once CNN or the major television networks cover a story, it is usually picked up by every other station, as well as by newspapers around the world. Most major cities today have only one newspaper, which has led to a change from reporters seeking a scoop to wanting to be the newspaper of record. Each newspaper follows the news agenda as set by the networks or CNN to ensure that they cover all major stories, so that they will then be available to sell later on the newspaper's web site (Adams 1998: 281). This "follow-the-leader structure" means that stories spread with incredible speed. For example, on 3 February 1996, Sergeant First Class Donald Dugan became the first US soldier to die during the NATO peacekeeping operation in Bosnia. He had picked up a landmine, which exploded. It was an accident, but CNN led with the story and that night every newscast led with the story. It also appeared on the front page of every major newspaper the next day. According to Lexis-Nexus, in the first 24 hours after his death, newspapers and magazines devoted more than 50,000 words to this single death: only a fallen general would have received such coverage in previous conflicts. CNN clearly set the agenda for the rest of the media and the story spread globally within 24 hours (Adams 1998: 278–9). A dramatic story based on altered images could spread just as quickly.

Contrary to popular wisdom, however, CNN does not, in most cases, set the agenda for the government. A study of US involvement in Somalia, for example, found that the government decided to intervene, which then led to – but was not in response to – media coverage (Bowdish 1999: 33–4). The government, especially the president, can set the agenda if it decides to, but if the government does not focus the media on one story, then the media will find another major story to cover. Of course, government scandals or wars can override even the government's ability to set the agenda. Especially at a time of relatively slow news, however, a bogus story based on altered images could gain widespread media attention.

The news media's focus on objectivity and covering "both" sides of a story also plays into the hands of a deceiver attempting to spread a story based on altered images. According to the news media, a story is almost never completely wrong. In the name of balance and objectivity, the proponent of a preposterous story must be given as much airtime and space as the rational critic of the story. For example, at the turn of the twenty-first century, almost every news-media organization covered the announcement that an obscure group was in the process of cloning a human. In the name of objectivity and the need to be newspapers of record, the news media covered the story en masse, but with little, if any, analysis of the validity of the group's highly questionable claim. The claim later proved false.

For a story based on altered images, such a system makes the deceiver appear equal to those seeking to expose and discredit the deception. This assumed duality, based on truth versus falsehood and drawn from such areas as the two-party political system and the adversarial legal system, fails when applied to many, if not most, news stories. Most stories have far more than just two sides. Furthermore, one accurate and one biased account, let alone two biased accounts, rarely leads to the truth winning out, especially since the media provides both views with equal time and rarely offers objective, rigorous analysis. Worse, television is usually viewed once, which provides no time to go back and analyze the information presented, let alone the images (Govier 1988: 45–50).

The emphasis today on spin also means that reality is lost amid the desire to shape perceptions with little, if any regard for the truth. Reporters are supposed to report reality, yet spin is regarded as an important part of that reality. Therefore, they report both reality and the false reality newsmakers are attempting to create. Reporters rarely comment on the difference between reality and the reality being offered by newsmakers because of the desire of reporters to be objective and not to appear biased (Kinsley 2005: M5). Both sides of a story, reality and spin, must be reported on equal footings, making the dissemination of a story based on altered images that much easier.

The meaning of knowledge has also changed, aiding those who seek to deceive. In the distant past, most of what people believed or knew to be true was based on their own knowledge and ability to reason and understand. Today, everyone believes many things for which they have no independent evidence. We often do, and must, rely on experts who know far more about a given topic than does the average individual, who may not even understand the expert's reasoning or evidence. Therefore, knowledge is no longer individualistic; it is communal. Even scientists rely on other scientists for information. In one physics journal, *Physical Review Letters*, for example, between 25 April and 18 July 1983, of 42 articles, 11 articles had more than 10 authors, 9 had more than 20 authors and 5 had more than 40 authors. Only five articles had one author (Hardwig 1988: 133, 137). Since 1983 such multi-authored research articles have only become more common in almost all branches of science as co-authors rely on each other to discover the truth in each of their small parts of a larger research project. For individuals, even in less esoteric topics, such as foreign policy and war, experts have now become the source of a great deal of not only information, but of judgments and conclusions.

Far worse, the media usually relies on very few experts in any given field. A handful of former secretaries of state, generals and professors dominate the media as experts on foreign affairs and war. For example, in February 1995 the IRA/British ceasefire collapsed after a bombing in London. James Adams reported the story and, based on his intelligence contacts, was then contacted by numerous other media as a major source for the story (Adams 1998: 282). The proliferation of transmitters of information (television, the Internet, newspapers, magazines and radio) has made it seem that the number of sources of information has increased, yet there has been a significant decrease in the number of sources used by the mass media (Rainey 2006: A7). A decline in the number of reporters, means that they have less time to find and interview sources, so they rely on the same sources repeatedly. They also rely on the same sources that have been cited elsewhere. Therefore, it is far easier to deceive the public if someone convinces, co-opts, bribes or manipulates just one of those opinion-leading experts into supporting a story based on altered images.

The proliferation of transmitters of information has also meant that the number of entry points where a manipulated image could be introduced into the media has exploded. Getting a story into the news stream becomes a numbers game. During the Carter administration, for example, the Soviets tried a new tactic. A US reporter would receive an anonymous telephone call from a supposed CIA employee stating that they had proof that Carter had recklessly planned the 1980 attempted rescue of US hostages in Tehran. *Washington Post* reporter George C. Wilson was contacted by just such a source, but sent the information he received to government sources before deciding whether to publish it. It turned out that the information contained many errors, such as incorrect code words and inaccurate information about the organizations involved in the rescue operation. Columnist Jack Anderson also received the information but published parts of it, prompting a firm denial from Jody Powell, Carter's press secretary. The Soviets had found a weakness that still exists today. If a faked story is sent to enough reporters someone inevitably will publish it (Brugioni 1999: 144).

Today, once a story that occurs outside the United States reaches the American media, there is far less chance that anyone with local experience and contacts will be available to verify its validity. This reality reflects a change in the structure of the news business. Until the 1970s, networks and large metropolitan newspapers stationed foreign correspondents in every major and many minor foreign capitals. In the 1980s, mass media became more of a business, owned by corporations, not by families and individuals with hard-news traditions. The corporate eye on the bottom line coupled with the rising cost of stationing correspondents overseas and advances in transportation, which meant that networks could speed correspondents to trouble spots around the world in a matter of hours at relatively low cost compared with stationing someone overseas full-time, led to a drastic decrease in the number of foreign correspondents. By the early 1990s, fewer than one quarter of the largest 100 US newspapers had a single full-time foreign correspondent (Rutten 2005: E6). Worse, the correspondents who survived are responsible for a larger number of countries, which decreases their expertise on, and ability to cover any one country. Therefore, local freelance journalists and stringers produce most of the day-to-day coverage of much of the non-Western world. A crisis brings in the big-name correspondents and sometimes the anchors, but even they rely on local stringers, reporters and guides as sources.

In one extreme case, Western reporters covering the fighting in Iraq after the 2003 invasion faced an extremely dangerous environment. In the first three years of the insurgency, 38 journalists were abducted and 82 were killed, including one Italian journalist who was beheaded (Rutten 2006: E16). As the insurgency worsened most Western reporters relied on Iraqi stringers to gather information and images about the fighting. Even Reuters, the largest news organization in the world, relied on Iraqis to gather the information and images that it then sold to media outlets around the world (Rubin 2005: A10). Consequently, the precise source of videotape, film and photographs from overseas is often unknown. Where was it shot? By whom? When? Most importantly, has it been altered? Often video is purchased from a third party with little analysis of any agenda the source might harbor (Amato 2000: Online). Worse, few, if any, journalists are trained to detect altered images. The system is admirably suited for adversaries to feed stories based on altered images into the global media stream through manipulated, threatened or bribed stringers or, in the case of Iraq, through moles who work for Western media organizations, but actually support the insurgents.

There has already been one known case of an altered video being fed to a Western news organization by a local source. A Palestinian cameraman supplied a French television station, France 2, with footage that was said to be of Israeli soldiers killing a 12-year-old boy on 30 September 2000. The video helped spark an uprising in the West Bank, but it appears that the scene was staged and that it is a case of a false voiceover. The video shows a man and boy crouched by a wall amid sounds of gunfire. A final burst kicks up dust that hides the pair. When the dust clears, the boy is lying down and a voiceover states that the boy is dead, although the dust obscures whatever happened. France 2 said that supporting footage of the boy's death throes was available, but later admitted that no such footage exists. The rest of the footage that is available, which was never shown on the station, show the boy getting up and dusting himself off. The voiceover said that the boy's name was Mohammed Dura and a boy with the same name did indeed die on 30 September in a Gaza hospital, but his face does not match the face of the boy in the video. Some analysts also have said that they have seen outtakes of the scene with Palestinians acting different roles. After five years, it appears that the truth has finally emerged, exposing the film as a hoax (Gelernter 2005: B11). Given the difficulty and amount of time involved in distinguishing truth from fiction in this case, it is highly probable that there are many other videos that have appeared on television that were staged but that have not been exposed.

As with foreign correspondents, there are also fewer reporters at newspapers and on television to fill even more space and airtime, because of 24-hour news on

CNN, MSNBC and other stations, and burgeoning newspaper websites. For example, in 1980 Philadelphia had 500 newspaper reporters, but by 2006 only 220 remained (Rainey 2006: A7). Therefore, most reporters today rarely conduct research. They have, for the most part, become stenographers of a sort, writing down quotes from witnesses of an event, with little, if any, analysis. Worse, the layers of fact-checkers, proofreaders, copy editors, editors, typesetters and printers who in the past helped verify a story, have been winnowed down by computers that put almost all of the editing functions into the hands of the reporter and, often, a single editor. Such a system has given rise to fake stories coming out of many major news organizations ranging from *The New York Times* to CBS News. The stories have often come from relatively low level reporters, demonstrating how easy it is to disseminate a fake story. Such deceptions are often not discovered until long after an event, and probably far more have occurred than have been exposed.

The dominance of images has also made it easier to disseminate invalid stories as propaganda or to deceive. In 1776, the printed word was dominant and was how most people received information. In the early 1800s, Alexis de Tocqueville wrote, "the politics of America was the politics of the printed page" (Postman 1988: 14–15). The nineteenth century, however, saw what Daniel Boorstin called the Graphics Revolution. Posters, photographs and drawings became dominant in advertisements and political discourse, as well as in wartime propaganda. The twentieth century witnessed the Electronic Revolution and images, especially on television and in motion pictures, became the dominant form of communicating information to mass audiences. Ironically, even the study of rhetoric itself has recognized this change from text to images and is shifting from studying words to studying images (Hill and Helmers 2004: ix).

Images, especially on television, now dominate our lives. Children watch an average of 5,000 hours of television even before they start school and 16,000 hours by the end of high school (Postman 1988: 12–13). The images vastly overshadow any accompanying text or words. In 1982, Richard Nixon was asked about Ted Kennedy's chances in the 1984 presidential election. Nixon said Kennedy would have to lose 40 pounds. Given the dominance of images over words, Nixon speculated that Americans would not elect a fat politician, no matter how eloquent (Postman 1988: 14). Michael Deaver, who handled President Reagan's public image, also understood that as long as the images of Reagan were upbeat and in patriotic settings, it did not matter what words accompanied them. A White House official once called Lesley Stahl of CBS News to thank her for a report on Reagan. Stahl was surprised, since she thought that it had been a critical report, but the official knew that the images were so pro-Reagan that it had not mattered what spoken words accompanied the images (Adams 1998: 289; Klein 2004a: 21). Exposition, explanation and argument, the traditional tools of rational discourse, are less and less parts of political discourse or foreign policy. Image is key.

Television news exemplifies this focus on images, not words. Television news is an entertainment show with 10–15 stories. No story is longer than 100 seconds and most are fewer than 45 seconds. The average length of a story on something as important as presidential elections, for example, on network evening news broadcasts fell from 42 seconds in 1968 to less than ten seconds in 1988 (Adams 1998: 280). There is little, if any context, explanation or theory, and little, if any, information. Once a story, whether bogus or valid, enters the news cycle, the brief time spent on each story, especially on television, means that there is little, if any, analysis of the content. There was extensive coverage of the Iran hostage crisis, for example, yet after the crisis, few, if any Americans knew what language Iranians speak, what Ayatollah means, Iranian's religious beliefs or much of anything about Iran, Iranians or the history and background of the crisis. Yet, everyone had an opinion, based almost entirely on images, not words. As Neil Postman (1988: 16–18) argued, these are, more accurately, emotions, not opinions. This argument explains why such "opinions," as surveyed by pollsters, change every week. They are emotions in response to images, not opinions based on a rational analysis of facts.

Even discussions on television, because of the quest for balance, become brief, almost unrelated sound bites on opposing viewpoints, with no argument, analysis or explanation. No one ever says, "What do you mean by that?" or "Let me think about that," because thinking does not play well on television. People do not appear to be doing anything if they are thinking. Therefore, it rarely even matters what is said, it just matters that someone is on television. Hence, Speaker of the House Thomas "Tip" O'Neill appears on the sitcom, *Cheers*, while presidential candidate Clinton appears on late night television playing a saxophone (Postman 1988: 16–18). Image is everything; words are nothing.

Because the media is now based on images, it already provides deceptive information or what intelligence agencies call disinformation: information that is not inaccurate, but which misleads. During the 1968 Tet Offensive in Vietnam, for example, images of VC in the US embassy in Saigon led many Americans to conclude that the war was lost. In reality, however, the offensive was a military disaster for the VC. The attack destroyed the indigenous VC force in the south (Summers 1984: 138, 183–4), but the images never conveyed that information. The focus was on US casualties and the temporary loss of part of the embassy. Such a focus is the result of the media's quest for ratings and readership, but the result distorts perceptions of reality. The focus on US casualties has continued today, with almost no media focus on enemy casualties, thus making even US battlefield victories sometimes appear to be defeats, such as in Somalia in 1993. In such cases, which are all too common, the information is fragmented, superficial and creates an illusion of knowing something, but does not provide valid information. It is the same type of information a psyops or deception operation using altered images would seek to disseminate.

Unfortunately, misinformation or inaccurate information often looks very much like accurate or valid information (Achenbach 1998: Online), making it far easier to disseminate a false story. Furthermore, inaccurate information, which tends to be more sensational than good information, often spreads faster than valid information. Although science, sooner or later, roots out invalid information, the media, individuals and society usually do not. In 2003, for example, some 60 percent of Americans surveyed believed at least one of three misconceptions about Iraq. The three invalid facts were that weapons of mass destruction (WMD) had been found in Iraq; that there was clear evidence that Hussein had worked with the 9/11 terrorists; and that people in other nations supported the war in Iraq or were closely divided in their support. In fact, no WMDs were found, Hussein had not collaborated with the 9/11 terrorists, and most nations opposed the war. Furthermore, of those states that supported the war, such as Spain, Pakistan and Britain, the majority of their populations, often overwhelmingly, opposed the war. More importantly, belief in this invalid information led to support for the war. While 83 percent of those who believed all three misperceptions supported the conflict, only 25 percent of those who did not believe any of the misperceptions supported the war. Slightly more than half (53 percent) of those who believed one misperception and 78 percent of those who believed two of the misperceptions supported the conflict, in a perfect correlation between belief in the misperceptions and support for the war (Davies 2003: 1). The key point is not whether the war was right or wrong, but that many people based their support for the war on inaccurate information that appeared indistinguishable from valid information.

Once a story based on altered images or video reaches the public, they are, for the most part, ill-equipped to judge whether it accurately portrays reality. It is an American belief going back to the Bill of Rights that a free marketplace of ideas will expose and discredit false stories, including those based on altered images, and that the truth will emerge triumphant. A free marketplace of ideas, however, presumes that the public has the ability to critically analyze information and arrive at well-reasoned conclusions. The use of images in the place of text to convey ideas has made this presumption far less valid than it once was, if it has any validity left at all.

The critical skills required to analyze information depend on the type of information, whether it is oral, written or image based. For example, if information is in the form of the printed word, a reader needs logic, rhetoric and semantics training to analyze it. Such education was stressed in eighteenth-century America when the printed word was dominant and speeches were vital to the nation's political life. Today, however, with images dominant, another set of critical skills is required, and that set is far harder to acquire than the set for analyzing text or oral rhetoric.

The major problem with appeals that involve images is that they do not make statements of fact that can be proven or disproven. To determine whether a claim is valid or truthful, it must be made in words in the form of a proposition or hypothesis, which affirms or denies a statement of fact. The difference between text and images is clear if a comparison is made between *The Wealth of Nations* and the Statue of Liberty. You can use facts to determine whether the arguments made in Smith's book are true or false. You cannot use facts to determine whether the Statue of Liberty, for example, is a valid expression of liberty. The statue evokes emotion, not logic. You can either like or dislike the statue; you cannot disprove it or say that Frederic Bartholdi, the sculptor was wrong. As with the Statue of Liberty, image-laden political (or commercial) advertisements, let alone propaganda, rarely, if ever, make statements of fact that can be tested. Political spots are a drama or a myth that can be liked or disliked but cannot be refuted because they usually do not make statements of fact. Unfortunately, news reports also are becoming more drama and less traditional news. Clear statements of fact are few and far between.

The dominance of images raises questions about the future of democracy. The Founding Fathers did not give the vote to young people because they reasoned that to be a good citizen, individuals needed a certain level of education, literacy and analytical skills to analyze political rhetoric and information in order to vote rationally. Based on the change in the way information is conveyed – from text to images – as early as 1954, George Counts at Columbia University argued that the media had repealed the Bill of Rights. This change had not been accomplished by limiting information, but by changing its form so as to make it extremely difficult, if not impossible, to analyze its validity and truthfulness. Even if information is readily available, if people lack the critical skills to evaluate it, "they are as vulnerable to tyrants and knaves as those who are deprived information" (Postman 1988: 12). As Postman argued, the irony is that whereas most people fear an Orwellian *1984* big brother that controls information, the future we have is the one portrayed by Aldous Huxley's *Brave New World*. No Big Brother is needed to deprive people of their freedoms. In our "Brave New World," most people embrace their oppression by loving the technologies that deprive them of their ability to reason. "Orwell feared we would be oppressed by tyranny. Huxley feared we would be oppressed by triviality. Orwell feared we will be controlled by fear. Huxley feared we will be controlled by pleasure" (Postman 1988: 18). Huxley appears to have been the more prescient futurist. The very love of images that makes truth so difficult to discern, also plays into the hands of deceivers who would alter images to deceive us or inundate us with propaganda that uses manipulated images. It is far more difficult, if not impossible, to argue against an altered image that arouses emotions than against an erroneous position stated in words in the form of a proposition.

All of these factors mean that the media does not, usually, help society determine which information is true and which is false. In many ways, in fact, the media facilitates the spread of disinformation, which can only help an adversary seeking to use altered images for propaganda and to support deception operations targeted against the United States.

Even though many of the industries related to the mass media and manipulating images are based in the United States and Americans see themselves as both technologically advanced and aware, the United States is extremely vulnerable to deception operations and propaganda based on altered images. Many other states have equally advanced technology to manipulate and disseminate altered images, albeit not in as many areas. India's Bollywood, for example, is as adept at producing motion pictures, albeit of a different type, as Hollywood, and Ireland and India, to name just two countries, have advanced computer software industries that rival, and in some areas surpass, that of the United States. Cell phone use is as common, if not more so, in Europe, Japan and Israel as in the United States, and robots are far more common in Japan than in America. Many other countries from the European Union to Japan and from Taiwan to Brazil also have strong advertising, mass media and public relations sectors that could easily be used to create propaganda and devise deception operations based on altered images.

IW, including psyops, propaganda and deception operations, is not a zero-sum game. Even if one country, such as the United States, is dominant, expertise in mass communications and computer technology related to manipulating images does little to defend against the use of altered images by others. For example, if a state has experts who excel at creating manipulated images, such skills do nothing to stop others from using altered images for propaganda or to support deception operations, except possibly by allowing the experts to identify altered images more quickly. Such identification, however, often can take even experts a long time and, even if altered images are identified, it can be difficult to convince the public that an image has been altered.

The best mark for deception is someone who does not believe that they can be deceived. Historically, powerful states often believe that they are too powerful, too technologically advanced, and have far too extensive intelligence networks to be deceived. The militarily weak, however, often deceive the strong, from the Trojans wheeling in their prized horse to Iraqi exiles deceiving the Bush administration into believing that Iraq was close to developing WMD in 2003.

The recent past is not reassuring for those who hope that US leaders do not share the hubris-based belief that the United States cannot be deceived. For example, Bobby Ray Inman, the first Director of the National Security Agency (NSA) and then, until 1982, Deputy Director of the CIA, said that US intelligence assumed that the Soviet data intercepted by satellites "came in relatively pure," untainted by deception. "The vastness of the (US) intelligence 'take' from the Soviet Union, and the pattern of continuity going back years, even decades… (makes Soviet deception) impossible" (Epstein 1989: 150). Inman was far from alone in this belief. "At a 1985 conference at the Naval Postgraduate School, CIA officials denied that they could be deceived" (Epstein 1989: 196). Admiral Stansfield Turner, head of the CIA until 1981, wrote in his biography that US sensors were capable of "seeing," "hearing" or otherwise "detecting" virtually everything of consequence that occurred in the Soviet Union. Such an assertion was, on its face, absurd. There were far too few satellites and far too much to see. The failure to foresee the Soviet invasion of Afghanistan, the fall of the Shah of Iran, and numerous terrorist attacks, such as those on 11 September 2001 belie the claim that the United States cannot be deceived. Satellites cannot see into buildings, let alone into the minds of an enemy, where deception plans reside (Epstein 1989: 151–6). Furthermore, even if satellites convey accurate information, that information can still be used to deceive by using it to reinforce an apparently valid, yet wrong conclusion an adversary wants US leaders to believe.

Some argue that unrivaled intelligence capabilities make it harder today than during the Cold War to deceive the United States. US military officers talk about "battlefield dominance," including information dominance, which strongly implies that they do not believe that they can be deceived. A 2004 NSA briefing boasted that the United States and its partners intercept more than 650 million "events," including radar signals, radio and data transmissions, satellite, cell and landline telephone calls, faxes, e-mails and text messages every day. The raw information, in dozens of languages, many encrypted, then goes to about a dozen centers around the world where computers automatically process it. From the mass of raw data, some 10,000 thematic and geographic "product reports" are created each day and sent to analysts at the CIA, NSA, military intelligence agencies and commands. In addition to this mountain of data are hundreds of thousands of reconnaissance photographs, human agent and law enforcement reports, prisoner debriefings and document translations, as well as information from open sources, such as radio and television broadcasts, newspapers, magazines and the Internet (Arkin 2004: M1).

There may be more chances to learn the truth from such a multitude of sources, but history suggests that the reverse is far more often the case. Ironically, such a vast intelligence gathering system provides thousands of avenues for disinformation, some of it based on altered images, to enter the US intelligence system. Furthermore, such avenues provide thousands of opportunities for feedback to let a deceiver know how his deception story is playing. In the past, intelligence services themselves often have been crucial conduits for disinformation. In fact, without the intelligence services, historically it would have been much harder to dupe the British, French, Soviets and the Americans, among others, since moles in their intelligence organizations provided feedback on deception operations run by their adversaries (Epstein 1989: 148–9).

Today, a mole might not even be needed to provide feedback on a deception operation. If the United States conducts a deception operation, it is often difficult to get feedback from the target about the operation's effectiveness, especially if it is directed at a closed, dictatorial state. Anyone attempting to deceive the United States, however, can rely on the media to report extensively on how a story is playing, sometimes even if it is supposed to be a government secret. The ship of state is one of the leakiest ships ever constructed and even secrets often become common knowledge. The media, therefore, can be a crucial element in a deception feedback loop. The KGB, for example, when they were running disinformation operations to blame the development of HIV/AIDS on the US military and to allege that Americans were buying Latin American babies for their organs (Andrew and Gordievsky 1990: 630–2) must have used the US media as a ready-made feedback mechanism on how well the stories were playing in the United States.

10

THE THREAT

Striking the media culture

Just as the militarily weaker have done for centuries, US adversaries will use deception, psyops and propaganda to attempt to level the playing field, both in the realm of diplomacy and on the battlefield. The modern capability to alter images will only make these methods, which historically have been remarkably effective, even more formidable. Although the United States and the West lead in the development and use of many of the technologies related to manipulating images, this advantage does little, if anything to prevent others from using altered images as propaganda and to support deception operations targeted against the United States and other media-saturated societies, such as Britain, Japan, Australia and India. Since psyops, propaganda and deception are time and situation dependent, it is difficult to devise specific scenarios to demonstrate how others might use altered images to harm US interests. Some general arguments, however, can be made and a few examples will provide a peak into the limitless realm of possibilities for such operations; a realm only limited by the creativity of the human mind.

The same attributes that made British and Americans good at deception and propaganda in the Second World War including high Mach scores, firsthand knowledge of the target state, mass communications experience and empathy for the target country, will make individuals adept at developing propaganda and crafting deception operations targeted against the United States today. French television correspondent Diego Buñuel is a reporter and is not a propagandist or deception operator, but his background illustrates the type of experiences those who oppose the United States would look for in an effective psyops or deception planner. He has mass communication experience. He works for the CAPA agency, which provides images for news organizations around the world. He also is familiar with US culture. Although born and based in Paris, he attended Northwestern University. When explaining how he filmed several exceptional pieces on US troops fighting in Iraq, he said;

> The freedom I get from our conversations [with US troops] is directly linked to my ability to understand American culture, to have lived it. … The US is a country where people recognize each other through

> baseball, football, TV series, nightly news, comedy hour, cartoons, magazines. These things form the recognition that I am dealing with someone from my tribe.
>
> (Hohenadel 2004: E25)

Many foreign governments hire American firms to conduct public relations for them in the United States. In 1990, for example, Kuwait hired Hill and Knowlton, a Washington, DC, firm to craft testimony for Congressional hearings about the invasion of Kuwait (Severin and Tankard 2001: 115). American Edward von Kloberg III made a career of polishing the images of dictators, including Saddam Hussein, Samuel Doe of Liberia, Nicolae Ceausescu of Romania and Burma's military regime (Washington Post 2005: B12).

White, gray or black propaganda that use altered images could be directed at the United States. Just as Fenton did in the Crimea, many governments use white propaganda to focus media attention on certain positive aspects of their country, while ignoring negative aspects. The Chinese, for example, attempt to focus attention on trade, while keeping the global media spotlight away from the brutal and repressive aspects of their dictatorial government. Similarly, many Arab states, such as Saudi Arabia, make it a point to educate some of their princes in Britain and America. The princes then serve as diplomats (and sources of white propaganda) in the West. They make their nation appear to be Western-oriented, even as their state remains fundamentally Islamic, with little or no democracy, strict media censorship and human rights abuses.

When gray propaganda is used, foreign news services can attempt to appear to be independent or quasi-independent, like the BBC, but really have their content tightly controlled by the government, just as the Soviet government controlled TASS. Such organizations can lend credence to stories based on altered images.

Finally, black propaganda can be spread by co-opted academics, such as Professor Segal who provided support for the Soviet HIV/AIDS story, or politicians and journalists, who may be in the pay of, be blackmailed by or be supporters of another nation or group. As the Soviets learned, if bogus information is sent to enough journalists, whether American or foreign, someone will publish it, regardless of the reliability of the source. Local stringers and journalists in other countries can also feed stories into the global news stream. In a similar fashion, seemingly independent and reputable-sounding front organizations can be created to feed stories based on altered images to the media. The Soviets, for example, created the African Friends Association to send distorted information on the US treatment of blacks to the media (Brugioni 1999: 142).

Successful propagandists will spread subtly slanted information or, even more effectively, selective truthful information that points to an invalid conclusion they prefer that reinforces an existing belief in a segment of the US population. Outright lies, by contrast, would not be nearly as effective. For example, an altered image could be of a laboratory that does conduct medical research, but what occurs inside can be made up, such as was the case when stories circulated that the US military created HIV at Fort Detrick.

An effective story based on altered images would not be of something verifiable or easy to refute, such as the sinking of a ship or an assassination. It is far too easy to show reporters that the ship is still afloat or that the leader is still alive. A successful altered image would probably be used to support a story that would involve few witnesses, such as an illicit affair, conspiracy or covert operation. By their nature, such stories would have few witnesses. Many people love conspiracies and such stories would find fertile ground not just in the United States, but also in target populations around the world that are already suspicious of the United States.

Timing is crucial to propaganda and deception operations: the key is to ensure that the target does not have time to disprove the story or trace it back to the source before the damage has been done. The use of altered images probably would be timed to influence public opinion or government decisions on time-dependent issues, such as during an election, just before a military intervention or during an international crisis. If, for example, the images of abused Iraqi prisoners had been staged by Iraqi insurgents and released just before a US election they could have had a significant effect on the results. As Kraig Short (Personal communication: 12 November 2005) suggested, such images released just before the State of the Union speech would force the president to respond to them, thus giving them a global audience well before they could be evaluated for authenticity.

In terms of themes, atrocity stories, whether based on altered images or not, are often effective and difficult to refute because in wars civilians are killed, hospitals are bombed and schools are destroyed. Furthermore, access to investigate such allegations in war zones is often limited, making the truth extremely difficult to establish, even long after the war is over. In 1973, for example, *The National Review* printed a story that said the bombing of Guernica during the Spanish Civil War was a Nationalist hoax, showing that the story, which was actually fabricated during the war, still had believers almost 40 years later (Knightley 1975: 206–8).

Altered images of atrocities could be used to attempt to incite a US or UN military intervention, just as images of dead babies from the *Lusitania* helped spark US intervention in the First World War. First published in 1915 to aid identification of the dead, such images were crucial to spreading the perception that the sinking was a barbarous act, even though the liner was listed in *Jane's Fighting Ships* and Brassey's *Naval Annual* as an auxiliary cruiser and was carrying a cargo of ammunition when she was sunk (Massie 2003: 534). More than 80 years later, images of atrocities in Bosnia sparked NATO intervention in the 1990s. In 1990 the Kuwaitis hired a public relations firm to help a young Kuwaiti woman make a presentation to Congress about barbarous Iraqi behavior during the invasion of Kuwait to ensure that the United States would intervene. Her testimony was extremely powerful but, as the furor resulting from the publication of images of US mistreatment of prisoners at Abu Graib prison in Iraq demonstrate, her testimony would have been even more powerful if she could have shown photographs or a video portraying Iraqi brutality. If it is well timed, the downside for such an operation is extremely small. Even when it became known that the woman was the Kuwaiti ambassador's daughter and that she was not in Kuwait at the time of the invasion, it was too late to change US policy, which supported Kuwait. The Emirate suffered little, if any, ill effects from the deception operation.

Images do not even have to be of corpses or genocide to change US policy. Images of US soldiers in 1994 standing by while Haitian police beat people celebrating the arrival of US troops led rapidly to changes in the US rules of engagement. In a similar fashion, manipulated images of brutality by a government allied with the United States, such as by Pakistan or Saudi Arabia, could be used to weaken US domestic support for US alliances.

Images of US casualties, whether chosen selectively, mislabeled, staged or altered to provide visual support for a bogus story, could be used in an attempt to change US policy. There is a widespread belief, reinforced by the US experience in Vietnam, Beirut (1982–4) and Somalia (1993), that the American public and their leaders cannot stomach very many, if any casualties, especially when shown on television. Although beheadings shown on the Internet do not appear to have changed US policy in Iraq, they might have had a significant effect if the beheadings had been of US soldiers. Such a tactic would risk infuriating Americans and lead to demands for revenge, but such images might also lead to calls for withdrawal. General Charles "Chuck" Horner, architect of the Coalition air campaign in the Gulf War, said "Public sensitivity to casualties can dominate our political and military decision-making in a crisis" (Adams 1998: 297). Such sensitivity is ripe for a clever propagandist to reinforce to end or even prevent a US military intervention. Adversaries might, for example, disseminate far more graphic images of US casualties than the US media is willing to broadcast or a video of an allied squad of soldiers in combat, such as a British unit, pleading for US air support and being told that none is available because the aircraft are supporting US units. If the British government said that the incident did not actually happen, many people would believe that such a denial was just the British government trying to maintain the Anglo-American alliance.

A propagandist could be less blatant than explicitly stating their opposition to an intervention by covertly funding a tour of images, a television program or even a motion picture that portrayed the costs, especially in individual terms, of the Vietnam War or the Beirut or Somalia interventions to decrease support for an imminent US military intervention. In fact, just broadcasting the actual words and images of US troops might have a powerful effect. In 2004, French journalist Diego Buñuel created powerful segments on the fighting in Iraq largely by letting US soldiers tell their own story. Many of the soldiers believed that the United States should not be in Iraq, sentiments rarely heard in the US media. "One said in my report," Buñuel recalled, "If I had troops in my hometown walking through my house, I'd probably pick up a gun and shoot them also" (Hohenadel 2004: E25).

Such information could be disseminated via the Internet to bypass US media gatekeepers who limit the negative information reaching the US public about military interventions and wars. Such self-censorship is common in every country and goes back at least as far as Fenton's images of the Crimean War. Images selectively chosen to highlight the brutality of war, such as of US troops bulldozing houses in an occupied area or, conversely, of US troops building schools, roads and bridges in a foreign land juxtaposed against images of decaying infrastructure in the United States might also weaken support for an intervention. Propaganda that reinforced basic, universal themes, such as the sanctity of life, that usually take precedence over national foreign policy goals, would also be powerful themes to reinforce in a campaign against a military intervention or a war.

Altered images in the form of staged events, mislabeled images and selectively focusing on certain things could be used to make a country appear far different to the US public and government than it is in reality. Images of free and fair elections, an independent press and of peaceful political demonstrations could be used to make a dictatorship appear democratic. In 2004 a group of Americans, including a former secretary of state, former US attorney general and an ex-presidential aide faced federal charges after they helped conduct a deception operation funded by oil companies that sought to make Kazakh President Nursultan Nazarbayev appear to be leading an emerging democracy. The plan succeeded in gaining the release of US aid that had been frozen because of corruption and human rights problems in Kazakh (Silverstein 2004: A1, A14). In what appears to have been a similar case in 2004, the government of Cameroon paid a team of former US Congress members to visit Cameroon to monitor elections. The team reported that the elections were fair, although other, nongovernment funded observers said the elections were "surrounded by fraud" (Silverstein 2005: A1).

Altered images, especially in the form of what the media selectively focuses on, can distort a nation's entire image. India, for example, is a democracy with a population that is relatively friendly to the United States, while Pakistan is a dictatorship with a population that is predominantly anti-American. The images of the two countries in the United States, however, are the reverse of reality. India is seen as anti-American, while Pakistan is perceived to be a friend. The Saudi image is also distorted, projecting a Westernized image to the West, obscuring an autocratic, monarchical state that many experts believe has leaders who support Islamic Fundamentalist terrorists.

Terrorism is another area in which altered images can and have already played a significant role. Images on the Internet of the spate of beheadings in Iraq in 2004 helped keep the war in the news, not to mention convincing some nations to reconsider their troop commitments or cut short their deployments. The Philippines withdrew its 51 troops from Iraq a month early in July 2004 to try to save one of its kidnapped citizens (Williams and Paddock 2004: A1, A5). Such beheadings, however, can be faked, as a San Francisco supervisory council candidate demonstrated in 2004 in a bid to attract attention. The Associated Press even reported on the alleged beheading of the man by an Islamic group (DiMassa and Hollis 2004: B4).

An enemy might create fake images of Americans executing prisoners of war. Such images would help convince their own troops to fight harder, increase hatred of Americans and decrease international support for the United States. In a more Machiavellian twist, one state or group might release images of an American hostage being tortured by a rival group or government to increase US opposition to, or even spark US military action against the rival group. In a similar fashion, Chechnya might release manipulated images of Russian troops torturing Chechens in order to decrease Western support for Russia's war in the region.

Terrorist groups might create entire programs to broadcast in the United States, probably via the Internet. If Al Qaeda announced that they were going to show a documentary on their inner workings, many curious Americans would watch. The program could acknowledge that the group has used extreme tactics, possibly even showing "inside" training and "combat" video from terrorist operations to attract viewers. (In fact, video of Al Qaeda attacks are common on their web sites.) The program, however, could also seek to reinforce some existing US beliefs by highlighting some things that Al Qaeda is fighting that most Americans also oppose, such as monarchies, big oil companies and nineteenth century borders drawn by imperial powers. Such tactics might lead some Americans to question US support for monarchies in Saudi Arabia and the Gulf States.

Propaganda based on altered images could be labeled "secret" to attract a larger audience. The Al Qaeda video mentioned earlier could be said to have been discovered during the search of a terrorist cave in Afghanistan and smuggled home by a US soldier, who then passed it to a journalist. Such a story would increase the size of the audience for such a video. Similarly, a video purporting to be of "secret" US military operations in North Korea might play well on US or foreign television. It would also present the United States with a difficult diplomatic crisis. Such images would provide the North Koreans with the moral high ground in charging the United States with violating their sovereignty and Washington would have difficulty refuting an operation that was allegedly secret. A similar operation could be conducted by the Cubans, Iranians, Russians, Chinese or the PLO to cast the United States in a negative light by claiming that Americans were conducting secret military operations on their territory.

For foreign audiences, images that reinforced themes of the United States as violent, racist, imperialistic, arrogant and politically naïve would be effective. US troops entering a mosque with their weapons and boots on, for example, with no explanation of why (such as searching for a sniper) or desecrating the Koran, for which stories have already circulated, would reinforce many Arabs' anti-American beliefs and incite Arab terrorist attacks against Americans. Images of US soldiers drinking beer at a base in the United States could be mislabeled as being a scene shot in Saudi Arabia. Worse, images of US soldiers could be altered to appear to be in Mecca or Medina, Islam's holiest places. Since the target audience would be the Arab states, the Western media might not even pay much attention to such a story – even discounting it as without any foundation – until it already had done its damage to the image of the United States overseas.

Altered photographs would play best in newspapers and magazines, while video would play best on television either as a news story, advertisement or an entire program. The Internet is also an excellent medium to spread propaganda since it has global reach, no gatekeepers and it is difficult to track stories back to their source. Other media in which altered images could be used are in books, which are often seen as more credible than a newspaper or magazine, and even in letters sent directly to opinion leaders in the United States and elsewhere. Even a motion picture can use altered images to present a distorted image of an event and, if criticized, those who produce such deceptive motion pictures can defend their propaganda by arguing that their movie is "just entertainment." For higher-tech distribution, DBS satellites can be rented or purchased and transmit programs directly into televisions in another state. The key, however, is to make the program entertaining so people will watch it, and to reinforce at least some existing beliefs so that it will not be dismissed as propaganda.

One major target of propaganda based on altered images would be carefully selected segments of the US public, but not the entire population. It is nearly impossible to change the opinions of true believers who oppose your cause, while preaching to those who already support you is a waste of propaganda effort. The goal, therefore, should be to target those segments of the population that are neutral or those who support a policy but do so with reservations. Even those who support an intervention, for example, could be targeted by highlighting underlying issues that worried them, such as casualties or the financial cost of the operation.

The other major target for propaganda would be foreign populations, especially those who are neutral or relatively pro-American, yet have concerns about US policy that could be reinforced. Propaganda also can put pressure on governments, especially in states where the government is pro-American, yet the people are anti-American. Countries such as Spain, Britain and Pakistan, for example, who had governments that supported the war in Iraq in 2003, but publics that did not, would be excellent targets for propaganda designed to increase popular pressure to stop government policies that supported the United States. Propaganda could ask why the government was pursuing a policy the majority of the population opposed.

Civic and independent political organizations also can be a target of propaganda. A story based on altered images could be started by sending images of atrocities, for example, to several human rights organizations, one of which might then disseminate them. This method has the added benefit of the organization then lending its credibility to the bogus story and masking the source of the images. Altered images could also be sent to non-Western media, which often are far more willing to believe ill of the United States than the US media. From there, Western media sometimes pick up the story or the target may be a foreign audience itself.

Just as with propaganda, altered images could play a role in any of the major types of deception operations against the United States, especially in the intelligence realm. As the capability to alter images makes it more difficult and time consuming to detect alterations, greater care must be given to determining whether images offered to support other intelligence information have been altered or actually do represent reality. One significant problem remaining from the past, however, is that even if an image itself has not been altered it may deceive by focusing selectively on certain things, be inaccurately captioned or may be of a staged event.

Just as the Serbs did against NATO in the 1990s, adversaries will continue to use camouflage and decoys against the United States, some of which may rely on altered images. The RAF used bomb craters painted on canvas to convince German pilots during the Blitz that airfields were damaged. Today, giant color images of bomb craters on a runway could be used to deceive photoreconnaissance satellites and aircraft. Furthermore, a ship could appear to be damaged after an attack by draping a billboard-style image of bomb damage over its hull. Conversely, a ship could be made to appear undamaged by laying an image of an intact hull over bomb damage to make an attack appear to have been a failure. The same technique could be used for military bases, buildings or any target. Such special effects can be extremely realistic. In 2004, for example, a motion picture company made a grain elevator in Baltimore appear to go up in flames, leading to clogged 911 lines, even though it was all movie magic and the structure remained undamaged (King 2004: E10).

Altered images could be used to support a spy's story about damaging a key building, just as the British in the Second World War used camouflage and decoy techniques to support their double agent, Zigzag's, story about damaging an aircraft factory. In a similar fashion, a state could make it appear that a nuclear reactor or chemical weapon factory was far from completion, when it was already operational. At the tactical level, life-sized images of guards, barriers and other defenses could be used to make it appear that a factory, base or transportation hub was better guarded than it actually is or could be used to confuse attackers as to which defenses are real and which are decoys.

Just as many states did in the 1930s, altered, staged or mislabeled images could be used to either make a state's armed forces appear more numerous and better armed or less numerous and more poorly armed than they are in reality. Pakistan, for example, might release altered images of more advanced Indian military forces to help convince the United States to sell Pakistan more advanced weaponry. Taiwan, Israel and many other states might try a similar deception operation based on mislabeled, staged or altered images of their adversaries' armed forces to convince Congress to approve the sale of higher-tech arms to them. US intelligence agencies might disagree but, if the images were released at a crucial time, such as when Congress was considering an arms package, the effect of the images might be felt before they were discredited. Such disinformation operations could be supported by other techniques, including information from double agents, propaganda, and diplomatic and economic measures, as well as relying for support on US politicians who already believed that the state in question was poorly armed compared with their adversary.

Altered images also could be staged to make a friendly military unit mimic an enemy unit. For example, North Korean troops dressed as South Koreans could be photographed or filmed in the Demilitarized Zone, allowing North Korea to blame South Korea for an incursion. Similarly, Palestinians could use their own forces to mimic Israeli troops desecrating mosques, harassing civilians or even just operating in areas of the West Bank or Gaza Strip from where Israel has withdrawn. Conversely, Israel could release staged video of Israeli agents mimicking suicide bombers who appear to be being coerced to undertake their missions by Palestinian militants. US troops also could be mimicked to film them desecrating mosques, torturing civilians or, more subtly, making derogatory comments about allied soldiers to try to convince other states to withdraw their troops from a joint intervention.

Doubles could be used to make it appear that US leaders or their top aides had met the enemy of an important ally in an attempt to weaken a US alliance. Anglo-American relations, for example, would be strained if images of a "double" President Bush or one of his top aides was seen meeting with representatives of the IRA. Conversely, doubles could be used to make it appear that a US ally was meeting with an enemy of the United States. What would have happened to the Anglo-American alliance if video of Prime Minister Tony Blair meeting with Hussein had appeared just before the war in Iraq in 2003?

Sometimes it is difficult to obtain photographs of wanted terrorists. In a form of dazzle, a terrorist organization could release dozens of images of a wanted man, with each image altered so as to slightly change the shape, size, color and texture of the suspect's features. Such an operation would make it extremely difficult to identify the suspect and would vastly increase confusion as similar looking men were arrested. Time and resources would be wasted. Conversely, hundreds of images of different people purporting to be terrorists could be fed to the United States via some of the thousands of conduits that feed information into the mammoth US intelligence community. US intelligence would be flooded with potential people for whom to watch for at US borders. Most of the images would not be of real terrorists, while the images of real terrorists would have been altered. If such a dazzle tactic was supported by rumors, forged documents, propaganda and double-agents, it would make the terrorists appear more numerous and powerful than they were in reality, as well as making US counter-terrorism measures less effective.

If hackers broke into government databases and altered the images of wanted terrorists, it would greatly increase the difficulty of capturing such people. Even if the alterations were detected, the United States would still question the authenticity of every terrorist description and image, just as Clarke did to the Germans by putting props under the wings of real RAF aircraft after the British realized their dummy aircraft could be identified from the air by the props under their wings. Furthermore, even just rumors that images of suspects had been altered would harm US credibility in prosecuting terrorists. Juries would wonder if the man in the dock had been correctly identified or if he was not the same individual as the government claimed him to be.

Operation Bodyguard dazzled the Germans with threats to Norway, the Balkans, the Pas de Calais and the south of France when the Allies were preparing to land at Normandy. A similar dazzle operation today might involve terrorists feeding images of multiple potential targets to the United States through the CIA, FBI or even the media. Terrorists could reconnoiter potential targets and then feed the images from the bogus reconnaissance to US authorities. They could make sure the Americans had to work hard to gather the intelligence to increase the likelihood that it would be believed. The images themselves would not have been altered, but they would suggest a deceptive reality in which terrorists were planning attacks on dozens of targets from cruise ships and shopping malls to water supplies and dams. Such a dazzle operation would harm various industries by scaring away customers, waste US resources, weaken defenses at real targets and increase the probability that the American public will grow tired of security alerts and largely ignore them, making future attacks easier and, probably, deadlier. The US response to bogus threats based on altered images also would allow terrorists to analyze the response and adapt future attacks to neutralize planned security measures.

Operations similar to the Soviet-run Trust in the 1920s and the Soviet operation that created bogus anti-Soviet rebels in Poland after the Second World War are tailor-made for the use of altered images. Attacks on police stations, assassinations and other anti-government activities could be staged or created using altered images to lend credence to the bona fides of a rebel group (really government operatives) seeking US support against an oppressive government. The goal would be to gather information on real government opponents and to harm US efforts to support real rebels. Similarly, video that is purported to be of protests against the government in North Korea sells for up to $200,000 in Japan, making it extremely tempting for someone to fake such images (Demick 2005: A1, A6). A US-supported government also could use altered images to make rebels appear stronger than they really are so as to seek increased US aid.

The Iran/Contra scandal is another case where images could have been used to support the existence of a notional moderate faction in a state opposed to the United States. Video of moderate conspirators meeting, especially if key government or military officials appeared to be present, or of crowds at anti-government rallies would go far to lending credence to the idea that a moderate faction existed. Such operations might cause US policy to be less stringent toward an adversary in order to allow "moderates" time to gain power, just as the British did for the supposedly anti-Nazi generals in 1939–40 and the Reagan administration did for Iranian "moderates" in the 1980s.

In another example, North Korea could stage a disinformation campaign regarding their development of nuclear weapons. Altered images of real Russian or Chinese weapons mislabeled as North Korean could go far in supporting a North Korean claim that they possessed nuclear weapons to trade for food or to stave off a US military attack, just as Hitler used exaggerated claims about his air force to deter a Western military response in the 1930s. Similarly, if Iraq had released images of nuclear, chemical or biological weapons with their delivery systems in its arsenal just before the United States intervened in 1990 or 2003 the planned US attacks probably would have been postponed as photo-interpreters analyzed the images to determine their validity. Such analysis can often take a long time and, even if they were shown to be invalid, US public support for intervention probably would plummet, since some still would not believe that the images had been altered. Such an operation would also have an unsettling effect on US allies.

Staged or altered images of US leaders meeting with an ally's domestic opponents, such as a meeting between US military officers and IRA, PLO or Hamas leaders could all spark embarrassing crises for the United States. This would be especially true if the foreign leader, such as the Israeli or British Prime Minister, was only shown the images and not permitted to keep them for what are said to be security reasons to protect an intelligence source. The leader then would have little chance to analyze the images' validity. Furthermore, such information might never even reach the United States, with Washington left to wonder why a previously friendly foreign leader suddenly became anti-American.

At an individual level, images could be altered to make it appear that a US soldier's wife was having an affair with a general. The information, provided to the soldier by a friendly foreigner, could help convince the soldier to use his position to gather secrets for a foreign power to punish the general by later exposing the lax security at his base and the loss of secrets. The soldier would be unlikely to be trained to detect an altered video, especially if it was supported by such easily counterfeited items as hotel and restaurant receipts planted in the wife's purse. As Kraig Short (Personal communication: 13 November 2005) suggested, the deception could be even more subtle by having a relative send a clipping from a hometown newspaper about a new bar opening to a soldier stationed overseas. The photograph in the story could show the soldier's wife getting friendly with another man. The image could be altered in a single copy of the newspaper or a photographer at a newspaper could be bribed, coerced or co-opted into altering an image or allowing the publication of a manipulated image. Such an operation also could involve a soldier receiving a clipping of a demonstration against an abortion clinic with an image of his wife being hustled inside. A friendly foreign operative who approached such a depressed or angry soldier would have a far better chance of convincing him to do something to betray his country than if such a deception operation had not been conducted.

Similar to US bombing warnings in the Pacific in the Second World War and during the Korean War, a terrorist group could release images of a possible target, such as a port, commuter rail station, stadium or a city's water supply to the media. "Terrorists" could be inserted into the images as they scouted the target. The story would be supported by rumors and information fed to US agents. Then, after no attack occurred, the terrorists could release images of another target, and then of another and another. In this "cry-wolf" scenario, the government's response, as well as the public's, would slowly decrease. Then, on the tenth or fifteenth "alert" a real attack would occur and the terrorists would be able to trumpet, especially to the rest of the world, that the United States could not even defend a target when they had been forewarned. Since the final real attack would receive more media attention than the false warnings before it, many people would focus on the American failure to respond effectively to the final warning, while ignoring the effect of the false warnings before it.

The same conditioning principle would work for a state planning an attack, just as Egypt did to Israel in 1973. For example, if China wanted to invade the Spratley Islands in the South China Sea, over which it claims sovereignty, the repeated release of images of Chinese troops embarking on landing ships to invade could lessen the US and Taiwanese response when an invasion was finally launched. The images could be altered, staged or simply mislabeled videos of earlier exercises. Such an operation could be supported by the actual movement of a small number of troops to harbors (mimicking a full-scale invasion force), rumors, propaganda, disinformation fed through double-agents and other indicators of an impending invasion. At the very least, the chance of achieving surprise would be significantly increased.

It is difficult to offer examples of how others might change the rules in the future through the use of altered images, since the possibilities are as limitless as the most bizarre ideas the British developed during the Second World War, ranging from using corpses to carry disinformation to the Germans to using an actor to impersonate Montgomery. Altered images could still, however, be used in a new way. For example, terrorists could film their attacks and auction the video off to the highest bidding news organization. Even the fact that such a video existed and was for sale would guarantee increased coverage and prolong the effect of any terrorist attack. Furthermore, the images could be altered to enhance the carnage or to make security personnel look incompetent or cowardly. In an effort to lower morale, video clips of dying US soldiers and Marines could be sent to their families with letters stating that the images were of their loved ones dying. Such a technique would be even more effective if the image purported to show the soldier or Marine abandoned by his comrades as he lay dying.

Photography can be used for propaganda and deception merely by focusing on certain things and not others, with no actual alteration of the images. Video of a terrorist training camp, for example, can show well-armed, uniformed young men rigorously training. The implicit message would be that the terrorists are well-supplied, dedicated and professional. By not showing many unarmed young men, guards keeping less than eager volunteers in the camp, old weapons and other signs that the terrorists are far from strong, the images would be deceptive.

Just as many states have done in the past, a country could use images of their best-equipped military units to present a powerful façade to deter an attack. Conversely, attention could be focused on a state's poorest equipped units to present an image to the world that they are no threat to anyone, especially just before the state plans to invade a neighbor. Similarly, part of a military unit could be shown in a video to be peacefully at its base, even while the majority of the unit was on its way to invade another nation.

Attacks and explosions are more dramatic than scenes of schools or businesses operating normally. Therefore, the majority of the coverage of military interventions is deceptive because of what is shown and what is ignored, even though journalists do not intend it to be so, because peaceful areas receive little media attention. Those who opposed a US military intervention could enhance this tendency by distributing images of local people demonstrating against the US presence, as well as heavily armed US forces interrogating civilians, putting black hoods on suspects, and shooting into what appear to be ordinary homes. Such coverage of an intervention would be distorted, since those welcoming and cooperating with US forces, and US troops rebuilding infrastructure or providing medical care would not be shown.

Photographs or video of a crowd filling the streets to listen to a politician gives the impression that the speaker is popular. By not showing that the crowd ends just outside the image's frame or the much larger demonstration for a rival candidate across town, the images are deceptive. The crowd that clustered around the fallen statue of Saddam Hussein in 2003, for example, appeared large, except for the few times when the camera pulled back to reveal that it was in reality rather small. Similarly, *Triumph of the Will* gave the impression that all Germans supported the Nazis by showing huge crowds at the Nuremberg rally. The film, of course, did not show the majority of Germans who did not support Hitler's party at the time it was made. Hitler's Nazi Party, in fact, never won a majority in an election, with its best showing in March 1933, when it won 43.9 percent of the vote.

Almost anyone can stage events to film or photograph to deceive or to serve as propaganda. Crowds can be hired, lured with free food and drink, or just coerced by the police or military into appearing at a political rally to be filmed or photographed. In 1953, for example, MI6 hired a Tehran weightlifting club to gather crowds for anti-Mossadegh marches in Iran to help overthrow the democratically elected leader. Terrorists can muster large numbers to appear at a training camp when, in reality, many of the "terrorists" are local people either hired or coerced to participate in the day's filming.

As in the motion picture, *Wag the Dog*, a small group of actors playing refugees could be filmed to appear as if thousands are present in an attempt to garner international aid, just as Emperor Selasie used the alleged Italian bombing of a Red Cross camp in Abyssinia in the 1930s to garner international support. The blame for the apparent exodus could be placed on an ethnic group, for example, to justify attacking that group or could be used by one state to cast blame on a rival state. Timing for such an operation would be crucial. If the story was said to be occurring in a location that was difficult to access, either because of a war or terrain, the story might have a long enough life to alter US policy or frame a crisis before the truth was discovered. Front organizations could be used to spread word of the disaster, while criticizing those who question the story as being in the pay of the aggressors.

Anti-American propaganda also can be staged. For example, journalists could be led to a destroyed mosque after pieces of a US cluster bomb have been added to the debris to cast blame on the United States for the building's destruction. Actors also could be dressed as US soldiers and filmed staging atrocities, such as torturing civilians or desecrating religious or archeological sites.

Altered images also could be fed into the feeds from UAVs and, possibly, satellites. In 1999 during NATO's Operation Allied Force, video processing technology was used to locate and target Serbian military vehicles in Kosovo. The project was called TIGER, targeting by image georegistration, which sought to determine the coordinates of an enemy vehicle and then launch an air-to-ground missile in real time. Georegistered images of Kosovo were overlaid onto the scenes coming in live from a video camera in a Predator UAV 450 meters above the battlefield. TIGER, which automatically detected moving objects, fed the coordinates of any Serbian vehicle in the Predator's view to targeting officers. The process resulted in 80–90 percent hits on mobile targets. The next step will be to hardwire TIGER to a fire system to attack anything that moves, similar to the Aegis air defense system on US warships that can be operated in the automatic mode to shoot down any incoming missile or aircraft. Predators are already used as unmanned combat air vehicle (UCAV) to launch Hellfire missiles at targets on the ground (Meyer 2006: A1, A28). In the future, adversaries probably will attempt to hack into the Predator's signals to alter the images the UAV transmits back to its base. Targets could be deleted or decoys inserted into the live feed. The same could be done to the feed from reconnaissance satellites. The transmissions are coded and scrambled, but nothing is unbreakable, as countless code-breaking successes throughout history illustrate.

Many nations have altered images to make their armed forces appear more numerous and powerful. Terrorists and states facing imminent US military intervention certainly might try the same tactic to discourage intervention, such as by showing that they have weapons of mass destruction when they do not. A foreign politician seeking to prove to the United States that they are popular may also forego the bother of actually having to stage a massive rally and just use computer alteration to make a video appear to be of a massive turnout for one of his speeches. The large crowd at a rival's speech could also be edited to make the opposition appear far less popular than it is in reality.

During a crisis, either side could alter satellite or reconnaissance images to either add or delete forces. If during the 1962 Cuban missile crisis, for example, the Cubans had provided images showing that there were no missiles on the island and "evidence" that the US images had been altered, it might have drastically decreased international support for the US naval quarantine of Cuba. It also would have been difficult for the United States to prove that their images had not been altered, since it is always difficult to prove that something has not occurred.

Altered images also could be used to make it appear that a politician or leader had met with unpopular or unsavory individuals, such as a convicted stock swindler, leader of a fringe political group, unpopular actor, terrorist, drug kingpin or dictator. Such a tactic would harm the leader's credibility, popularity and, if released during an election, could harm their chances of winning, as was done to Upton Sinclair in 1934, and Tydings and Douglas in the 1950 Senate elections. Such tactics are still used, as the attempt to use an altered image to link John Kerry and Jane Fonda during the 2004 Democratic presidential primaries demonstrated (Glionna 2004: A24). Such tactics also could have international repercussions. Images of a secret meeting between a high US official and leaders of a state the United States was attempting to build a coalition against, such as Iraq in 1990 or 2003, would weaken the US position. Potential allies would wonder if the United States really was intent on taking a hard line against the state in question.

Captions also can be changed. Images of a palatial home and expensive sports cars can be mislabeled as the home and cars of a leader of a poor state allied with the United States. If the leader has been trying to appear to be a humble man of the people, such propaganda would be damaging, especially if such images were released shortly before an election so there would be little or no time to refute the charges.

Images of civilian casualties or of bombed hospitals, schools or religious sites can be said in captions or a voiceover to be caused by an adversary, regardless of who is responsible and regardless of whether the images are actually from the current conflict. Images of weapons also can be labeled as belonging to whoever a deceiver wants. If the IRA was supposed to have disarmed after an agreement, for example, old images of armed IRA members could be relabeled and used to show the organization had not actually done so. The dates of a meeting between leaders can also be altered. For example, if a leader has fallen from grace, an earlier meeting between that disgraced leader and a new leader could be used, by changing the date of the meeting, to harm the new leader's reputation. Furthermore, as the Soviets proved with film of their ABM tests, video of weapon tests can be made to appear successful or to have failed, depending on the voiceover or supporting story, regardless of the actual effectiveness of the weapon.

A photomontage does not alter images, but juxtaposes them for maximum effect. As propaganda on their web sites, Al Qaeda already uses photomontages of young Arabs at training camps intercut with images of the airliners hitting the World Trade Center on 11 September 2001, as well as images from other terrorist attacks. As with the East German film about US pilots during the Vietnam War, interviews with smiling US pilots, said to be just back from missions over Iraq during the war, could be intercut with images of dead and horribly wounded civilians or destroyed hospitals, schools and mosques. Such photomontages could have a strong effect in much of the world by reinforcing existing anti-American attitudes.

Images that contradict a leader's words can also be powerful. Images of US troops at bases around the world could be intercut with statements by a US president that America seeks no empire and just wishes to be at peace and left alone. For a US audience, video of crumbling infrastructure in the United States juxtaposed with US troops building new roads, bridges and schools in another country could erode support for a foreign intervention. Photomontages also could be used to weaken an alliance. Images of the US President and the Saudi King meeting and proclaiming their friendship intercut with images of Saudi beheadings, religious police harassing women and of mistreated foreign domestic servants in Riyadh would weaken the US public's support for the US–Saudi alliance.

In all their forms, psyops, propaganda and deception operations are well suited to the use of altered images. Although the United States has the technology to manipulate images and has done so in the past, the United States is also a prime target for such operations and has already been targeted. New technology to alter images will only increase the effectiveness of such operations, which have been extremely effective historically. The threat from the use of altered images is real and the United States must be prepared for such attacks or risk losing the war of perception for the hearts, minds and eyes of those involved in foreign crises, military interventions and wars.

11

DEFENSE

The media culture strikes back

The United States must be prepared to defend itself against the use of altered images as propaganda and to support deception operations by other states, organizations and individuals. However, as the Biblical motto of Mossad, Israel's intelligence agency, warns, "without subterfuge, the nation falls," the United States should not cede to others a monopoly on the use of such deceptive stratagems. In 2004, the September 11 Commission urged the United States "to engage more deeply in a 'struggle of ideas' against Islamic fundamentalism" (Richter 2004: A1) and altered images can be powerful weapons in any "struggle of ideas" when they are used in propaganda or to support deception operations. Training personnel to use altered images for propaganda and to deceive will create a pool of experts who can use such techniques, as well as help detect altered images when they are used against the United States and help devise countermeasures. In considering how the United States might use altered images, however, legal, ethical and political constraints must be considered first.

US laws have not caught up with the use of information as a weapon, let alone the use of manipulated images. Legally, the US government can lie to the American people, as well as to foreign audiences. It would be extremely rare, however, for an administration to take a narrow, legal view of lying either with words or altered images. The American people, especially since Vietnam and Watergate, can become extremely angry with an administration that lies, especially about foreign policy or war. Furthermore, although an administration may seek to distort the truth, other branches of the government often will oppose such lying, such as when Congress demanded access to Nixon's tapes to expose his lies about the Watergate break-in.

Whatever the positions of particular administrations, the US military opposes altering images, at least when they are for the US public or military. A 29 August 1995 Directive on the Alteration of Official Department of Defense (DOD) Imagery largely forbids any alteration to images. Cropping and sizing, and some wet darkroom techniques were allowed, but only if they do not alter the "truth" of the photograph (Brack 1996: 49). The directive was issued so that not only the public could have faith in DOD images but so that military planners can trust the imagery they use to plan operations. Reconnaissance teams today often use digital cameras to take photographs, which are then printed and distributed to commanders to plan combat operations (Brack 1996: 48).

The target of propaganda or a deception operation is crucial in determining its legality and potential political costs. For example, although the US military bars manipulation of official images, US Joint Publication 3-58 states that US forces may employ deception of all kinds against hostile forces, with no legal or ethical constraints. Such deception would include the use of altered images. For example, an investigation found that the Pentagon's clandestine planting of stories in Iraqi newspapers did not violate US law or military guidelines (Mazzetti 2006: A3). By contrast, it is against policy to deliberately misinform or mislead the US public or decision-makers, except in cases where operational security is involved. Such a neat differentiation, however, ignores the fact that information that is directed at foreign audiences often feeds back to the American public via the global news media and the Internet. Besides the unintended audience problem, the issue of the use of deception against neutrals, noncombatants and adversaries with whom the United States is not officially at war is left open, as is a clear definition of "operational security" (Gerwehr and Glen 2000: 35–6). Because of such complex issues, most psyops, especially at the strategic level, currently require the approval of the National Command Authority, basically meaning the President (Bowdish 1999: 36).

Gray areas and presidential approval aside, most US military officers believe that they have significant leeway in their use of psyops and deception. In 1995, for example, the Freedom Forum First Amendment Center found that 60 percent of 1,000 US officers surveyed agreed that "military leaders should be allowed to use the news media to deceive the enemy, thereby deceiving the American public" (Adams 1998: 284–5). The military and politicians, however, should be wary of using psyops and deception that mislead the public because even plans to deceive foreign media can cause political fallout. In 2002 the Bush administration created the Office of Strategic Influence at the Pentagon, but quickly closed it after reports surfaced that the office planned to plant false news stories in the international media (Mazzetti 2004a: A1). Even training for psyops and deception operations can raise concerns among the public. Training in modern media technology and methods is crucial to developing effective psyops, yet concerns were raised when the Fourth Psychological Operations Group based at Fort Bragg, NC, had personnel training as interns at CNN's Atlanta headquarters. Critics charged that the military personnel were involved in news production, although CNN denied the accusation (Cockburn 2000: Online). Whatever the validity of the charge, it shows the level of public concern with keeping the government and media distinctly separate.

Such public concern severely constrains the US government and military from lying to the American public. The credibility cost of outright lies is usually far higher than the benefits gained through the use of such tactics. As the British discovered in the Second World War, the seemingly negative constraint of seeking to tell the truth can pay substantial dividends in terms of credibility, and the effectiveness of propaganda and deception operations. Truthful propaganda is usually credible and effective, while deception operations that use the truth to reinforce existing beliefs in order to reach an invalid conclusion are far more effective than attempts to deceive using lies.

The decision of whether to lie is complex because there are different degrees of lying and deception. One key difference is between keeping a secret, especially to maintain the operational security of a military operation, and telling a lie, which is far worse in terms of sacrificing credibility. There is also a difference between a lie of omission, by not providing all of the available information to the public, and a lie of commission, when the truth is knowingly distorted. The latter harms credibility far worse than the former. Given credibility concerns and the goal of an open, democratic government, the starting point for the United States should be not to lie. The burden of proof should be on those advocating the use of deceit to justify its use. A serious problem is that once an administration starts lying, especially to its own people, they are likely to start believing their own lies, gravely harming decision-making. Such constraints, however, far from rule out the use of all types of altered images in all cases.

The higher the stakes in a war or conflict, the more the military or an intelligence organization is allowed to do. If the future of a state hangs in the balance, as Britain's did in 1940, then almost anything is allowed no matter how brutal or immoral. British General Gibbins of the Special Operations Executive (SOE) called the secret war in the Second World War "bloodier than the Somme" (Brown 1975: 191). Colonel William H. Baumer, an American member of the LCS, later wrote:

> [Colonel] Bevan [LCS Chief] and his boys were extraordinarily clever. They knew exactly when and where to play upon Hitler's fury, and they did so knowing full well that if what they did became public property in later years they might all earn public opprobrium. But they were quite academic about this at the time. They were playing for the highest stakes imaginable, and there was no time for squeamishness. They knew they simply could not lose because, if they did, they and Britain would lose everything.... It was a dangerous business for they never knew from minute to minute whether they might be exposed and, thereby, expose the very matters that they were seeking to hide.
>
> (Brown 1975: 8–9)

After the war, Bevan said, "I do not think I should say what I did. I do not think Governments should admit to such matters, even if they were done in wartime" (Brown 1975: 270).

In a few cases Anglo-American deception and psyops in the Second World War even led to sacrificing American and British lives. De Guigand recalled, "On at least one occasion, the deception people were authorized to reveal the target of a major air attack on a German city to the Germans beforehand in order to reinforce the credibility of an agent who was to be used to mislead the German high command during Neptune [Overlord's early name]" (Brown 1975: 501–2). Alerting the German defenses before an Allied air attack undoubtedly cost more Allied airmen their lives than if the Germans had not been forewarned.

The British and the Americans also used propaganda to lure the Luftwaffe into the skies before D-Day so the Germans could be shot down, probably costing the lives of many Allied bomber crews. The Luftwaffe had been husbanding its aircraft to oppose the invasion, but Allied propaganda asked, "Where is the Luftwaffe?" German soldiers were fighting in Italy and the Soviet Union, while U-boats were fighting the brutal battle of the Atlantic, but the Luftwaffe was nowhere to be seen (Carroll 1958: 373–80). Stung by the challenge, the Luftwaffe came up to fight in force and the Eighth Air Force suffered losses of nearly 10,000 men, 800 bombers and 500 fighters in the three months before D-Day. Added to the casualties of the other Allied air forces, the losses almost doubled during the period compared with the previous three months, but the badly mauled Luftwaffe did not influence the Normandy landings (Brown 1975: 502). Eisenhower summed up the issue when he said, "One of the duties of a General is to determine the best investment of human lives. If he thinks the expenditure of 10,000 lives in the current battle will save 20,000 lives later, it is up to him to do it" (Brown 1975: 501).

Other states at other times have also employed what seemed like to some morally questionable tactics. In 1948 when five Arab armies attacked Israel and the country's survival was at stake, Israel even used torture, albeit for only a brief time. When Prime Minister and Defense Minister David Ben Gurion found out, he was furious and immediately ended the practice (Raviv and Melman 1990: 5, 24). Later, in 1956, Israeli counter-terror assassinations began and included the killing of those involved in the 1972 Olympic massacre in Munich (Raviv and Melman 1990: 122). Even during peacetime when the stakes seem lower, an Israeli agent underwent an operation to reverse circumcision, so if he was captured he would not be revealed to be Jewish. Conversely, an Egyptian agent was circumcised so as to pass as an Israeli Jew (Raviv and Melman 1990: 56, 152). Agents often use sex during times of both war and peace to lure foreigners into betraying their countries (Raviv and Melman 1990: 93, 105, 134). Such measures suggest that the use of altered images to support deception operations in the intelligence realm would be acceptable at any time, even if the public later found out about their use.

In terms of propaganda and deception operations that have public facets, both the American and British governments have lied to their own people, which suggests a willingness to use altered images even if their own citizens would be adversely affected. They have often lied via the media and, sometimes, with the media's cooperation. Usually such lies have related to the operational security of military operations, which is more a case of keeping a secret than telling a lie, and have not been designed to keep a government in power, as is frequently the case with the propaganda and deception operations of many dictatorial governments.

As with deception operations, the stakes determine how much, and the type of lies a democratic government will tell its own people through its propaganda. As J.C. Masterman, a top British deception officer, wrote after the Second World War, "The greater the advantage the greater the price" governments have been willing to pay to achieve it (Brown 1975: 404).

During the Spanish–American War, for example, a strong Spanish squadron was on its way from Spain to Manila to fight Commodore George Dewey's weaker squadron. The US government spread a rumor that an American squadron was on its way to attack Spain. The rumor reached the Spanish government, the Spanish squadron was recalled to defend Spain and Dewey's ships were victorious over their weaker foe in the Battle of Manila Bay. The rumor was, of course, false, but since it helped lead to a victory, the American public did not seem very concerned with having been misled (Uhlig 1994: 63).

Every government involved in the First World War lied, both to their own people and to foreign audiences. All governments lied about casualty figures (Knightley 1975: 108–9) and there were so many atrocity stories, often disseminated by the British (MI5 and MI6) and the *Bureau de la Presse*, which was funded by the French Secret Service, that the French press stopped printing individual stories and just ran them under the heading, *Les Atrocites Allemandes*. Few, if any of the atrocity stories on either side were substantiated after the war. All British propaganda files and records, however, were deliberately destroyed after the armistice to avoid exposing the government's role in creating and spreading such stories (Knightley 1975: 80–4).

Even during peacetime, democratic governments have disseminated far from accurate propaganda through subsidized friendly newspapers, as well as through clandestinely paid foreign journalists. A 1913 French parliamentary report called the need for such subsidies "incontestable" (Andrew and Gordievsky 1990: 31–2). Such subsidies continue today in many nations and even some democracies, such as in Mexico, where candidates and officials often pay for favorable news coverage (Kraul 2004: A1).

During the Second World War, the BBC and British newspapers bowed to the will of the LCS for deception operations, including accepting heavy censorship (lying by omission) and sometimes misleading the British people (lying by commission). Before the D-Day landings, the Allies leaked fake stories to the press to make FUSAG in East Anglia and the threat to the Pas de Calais appear real. The press reported that vicars were complaining about the "moral collapse" that had occurred in East Anglia since the arrival of the "vast number of foreign troops." There was also discussion in the press about "the immense numbers of rubber contraceptives" found around "the American paratroops' bases at Marham and Coggeshall" (Brown 1975: 604–5). Stories about "4th Army football matches," marriage announcements for men from notional units and music on the radio dedicated to phantom units helped reinforce the deception about FUSAG and the Fourth Army in Scotland. In 1944, the Allies fed rumors to the press about the British collecting information about Scandinavia related to a possible invasion

(Brown 1975: 466–7, 480). The press also reported that rehearsal exercises for the Normandy landing were the first of such training exercises, not the later stages of them, thereby masking the timing of the invasion to mislead the Germans. The deception, of course, also misled the Allied public (Brown 1975: 545).

The LCS even controlled what politicians and clergy publicly said about the Normandy invasion. The LCS recommended that the Archbishop of Canterbury not announce a national day of prayer on D-Day because it would tell the Germans that the Normandy landing was the major invasion (Brown 1975: 435). The LCS also wrote speeches for key Allied leaders ranging from Eisenhower and Roosevelt to Churchill and King Haakon of Norway to deliver on the morning of the invasion. To confuse the Germans about the true nature of the landing, the LCS speeches suggested that the Normandy landing was just the initial landing in a series of landings (Brown 1975: 671–2). In Parliament on 6 June, for example, Churchill spoke only at noon. Then, after speaking for 10 minutes about the liberation of Rome on 5 June, he paused and, following LCS guidelines, said, "I have also to announce to the House that during the night and the early hours of this morning the first of a series of landings in force on the European Continent has taken place… In this case, the liberating assault fell on the coast of France." The Germans were meant to assume other landings would follow elsewhere, but the public in Allied states was just as likely to be misled (Brown 1975: 675).

De Gaulle was the only leader to ignore the LCS script. He called for a national uprising, which the LCS and Allies did not want (Brown 1975: 675–6). The Political Warfare Executive (PWE), however, fed a story through an agent to the Germans that the Minister of Information, Brendan Bracken, had attempted to dissuade Churchill from mentioning that the landing was the first of many in his speech before Parliament. De Gaulle's statements appeared to support the story that Churchill had not followed the plan, while de Gaulle had. Therefore, in this double ruse, the Germans were meant to think that there would be other landings.

The Pacific theater also saw the US government use the media to deceive the Japanese, which in turn meant misleading the American people. In the summer of 1944, the Americans staged Operation Wedlock to deceive the Japanese into believing that the Americans would attack the Kuriles, when, in fact, Admiral Chester Nimitz planned to attack the Marianas. The press was encouraged to discuss the advantages of the shorter, northern great-circle route to Tokyo and to speculate about what was going on in Alaska, where dummy airfields and landing ships were constructed on Attu. The bogus story worked. The Japanese increased their Kurile defenses from 25,000 men and 38 planes to 70,000 men and 589 planes (Holt 1994: 48–9, 56–7).

More recently, when the stakes were arguably lower, the British government and military performed a deception operation involving the media during the 1982 Falklands War. The British media gave the impression that the Royal Navy planned either a frontal assault on Port Stanley, hit and run raids or a continuation of the naval blockade. This media deception helped the British surprise the Argentines when they instead landed at Port San Carlos (Thompson 1985: 54–78).

The United States has also used the media to disseminate deceptive stories. Four months after a US bombing raid on Libya, the *Wall Street Journal* on 25 August 1986 reported rumors that Qaddafi was planning new terrorist attacks. The newspaper also reported that the Pentagon was readying new bombing plans so as to be ready if the President ordered such an operation. No such attacks occurred and it later turned out that the story may have been leaked by the National Security Agency as part of a deception operation designed to encourage dissidents to attempt to overthrow Qaddafi (Goldberg 1989: 161–2).

During the Gulf War, the CIA and MI6 both had active disinformation campaigns targeting the Arab press to spread negative stories about Saddam Hussein. Both organizations, however, were careful not to manipulate the American or British media, although the United States and Britain both sought to aim the media searchlight at certain aspects of their military preparations, while concealing other aspects (Adams 1998: 286). During Operation Desert Shield, the US military focused reporters' attention on the arrival in Saudi Arabia of the 82nd Airborne, F-15C Eagles and the Marines, while downplaying the lack of armored forces that would leave the Marines and air-mobile units dangerously exposed for weeks. The goal was to make Hussein believe that the United States had sufficient forces in the Kingdom to defend it effectively against an Iraqi invasion, when, in reality, it probably did not (Hallion 1992: 135–8).

Just as Fenton had done during the Crimean War, during the Gulf War the US military focused attention on certain positive things and not on other, less savory aspects of the war, such as the US Army's use of bulldozer blades on M-1A1 Abrams tanks to bury Iraqi soldiers alive in their trenches. In an example of a more deceptive altered image and a form of false captioning, during one of General Norman Schwartzkopf's televised briefings, he nicknamed a man crossing a bridge in a vehicle, "the luckiest man in Iraq." The bridge is then hit by a bomb and is destroyed. What the general did not mention was that the blast from the explosion most likely killed "the luckiest man in Iraq."

The US military's presentation of video of missiles and bombs hitting individual buildings also negated any later attempt by the Iraqis to suggest that the Americans were hitting civilian targets or doing extensive collateral damage, regardless of the findings of any later studies. The image was firmly set in the minds of the public, especially in the United States, that US weapons were so accurate that any civilian casualties were merely unfortunate accidents. The military, in fact, had used carefully selected images to deceive. Even though most of the precision-guided munitions or "smart" bombs probably hit their targets, the vast majority of the bombs used during the Gulf War were unguided or "dumb" weapons that were far less accurate. Attacks with dumb bombs, however, were not shown to the media. Furthermore, even during the fighting in 2003, more than a decade later, some precision-guided munitions missed Iraq entirely and hit Iran, Saudi Arabia and Turkey. Such accuracy is far from the precision image crafted by the military during the Gulf War. Once the images of a war are set in the public mind early in a conflict, however, they rarely change. As columnist John Leo wrote, "Facts now have to play catch up with the images – and rarely win" (Adams 1998: 289). At least as far back as the deceptive images of the apparently highly mechanized German army racing across Poland and then France, being first is crucial to establishing the public image of a war.

NATO has also altered images to mislead the press and the public. In 1999, NATO accidentally bombed a train crossing a bridge at Gurdulice in Serbia. The bridge was the target, but the train happened to be on it at the crucial moment. NATO released a film showing the train coming into the bombsight, but too fast for the pilot to abort. It was convincing evidence that the pilot could not have avoided hitting the onrushing train. The film was, however, deceptive; it had been speeded up (Latimer 2001: 295).

More recently, on 14 October 2004, a US Marine spokesman in Iraq announced, "The troops have crossed the line of departure." The statement was taken to mean that the attack on rebel-held Fallouja had begun and CNN reported that the attack had started. In fact, the statement was a lie and was intended to allow US commanders to see how guerrillas in the city would react if they thought US troops were attacking (Mazzetti 2004a: A1).

When the situation has been serious enough, even during a military intervention and not a full-scale war, the United States and other democratic governments, including Britain, have sometimes misled their own people. There have been many cases of outright lies, both of omission and commission. Such lies have usually been used to gain a military advantage and have caused little, if any, negative political fallout. The United States and other democratic governments have also used certain types of altered images, such as focusing on certain things and not others, altering film and using false captioning, as well as photomontages, which are routinely used in domestic political advertisements. It would be erroneous to think that the deception operations of the Second World War would not be conducted today because of changes in the relationship between the media and the public on one hand, and the government and the military on the other, especially since Watergate and Vietnam. If the situation was perceived to be dire enough, such deceptions would certainly be repeated. As the history of various scandals shows, governments have often tried to deceive their public even when the stakes have been far lower.

The United States already produces vast amounts of white and gray propaganda that is created by the military and all the civilian branches of government. The United States probably also produces black propaganda, but only a select few know about it. US propaganda, however, often lacks focus, is seen as propaganda, and is general, not specific to a particular crisis. The United States is often hesitant to use its psyops capabilities during crises when psyops usually take a back seat to diplomacy, economic measures or military solutions based on the use of force. For example, the United States did little to counter the propaganda that was instrumental to the 1994 genocidal killings in Rwanda. When the plane carrying Rwandan President Juvenal Habyarimana crashed on 6 April 1994, Radio Television Libre des Milles Collines (RTLM) or Free Radio-Television of the

Thousand Hills, accused the Tutsi-led Patriotic Front of shooting down the plane. Before the Front could gain control of the country after overthrowing the Hutu regime, Hutu extremists used a mobile radio to incite their people to a mass killing spree of Tutsis, even as they themselves fled to camps in Zaire, Tanzania and Burundi. Radio announcers broadcast:

> Take your spears, clubs, guns, swords, stones, everything. Sharpen them, hack them, those enemies, those cockroaches... Hunt out the Tutsi. Who will fill up the empty graves? There is no way the rebels should find alive any of the people they claim as their own.
>
> (Bowdish 1999: 36)

The station was dubbed Radio Hate and the Hutus slaughtered the Tutsis. Even when the Hutus were in camps in neighboring countries, the same Hutu leaders who had incited the killing of the Tutsis used RTLM to order Hutu refugees to remain in the cholera-infested camps and not to return to Rwanda for fear of Tutsi retaliation. The Hutus stayed in the camps and some 3,000 people a day died from waterborne diseases (Adams 1998: 272–3).

The case highlights what the United States and the West could have done to prevent or alleviate the disaster. Major General Michael Hayden, part of the planning team for a US Joint Task Force (JTF) sent to Rwanda and Zaire to ease the suffering, later said, "We had psyops units in the JTF so we could have counter-punched, but we didn't. We are a rich and powerful nation, we had the means to find out where the radio station was, but we did not" (Adams 1998: 274). The head of the UN force in Rwanda, UNAMIR, Canadian Major General R.A. Dallaire, later wrote:

> These broadcasts were largely responsible for spreading panic that, in turn, drove large numbers of people to refugee camps in neighboring states, thereby spreading instability throughout the region.... It should also be pointed out that the broadcasts discouraged survivors from returning to their homes, and should have been jammed. The United Nations should have aired counter-broadcasts to give the population a clear account of what was actually happening... yet... no country came forward to offer jamming or broadcasting assets.
>
> (Adams 1998: 274)

The radio station was, in fact, broadcasting from the back of a Toyota truck. One special forces strike would have destroyed it, but such a military mission seems to have been perceived to be out of step with the peacekeeping mission.

In Bosnia and Croatia in the 1990s, radio stations also broadcast propaganda that aroused ethnic animosities and hatreds. A reporter in the area asked the US psyops team operating a NATO radio station, "Would the station run news bulletins to contradict the garbage pumped out by Serb, Croat and Muslim stations? A clear interpretation of the Dayton plan? Information about war crimes? No, the officers replied. The station will put out an anti-mine awareness campaign and 'good old American rock and roll'. Said the colonel: 'We don't get involved in politics' " (Adams 1998: 90). Finally, in October 1997 Stabilization Force (SFOR) troops shut down four television transmitters that allowed Serbs free access to the airwaves, dealing Serb propaganda a serious blow.

The United States could use its extensive media and mass communications capabilities against its enemies, as well as against those who incite genocide and war. Motion picture, television, radio, public relations, Internet and advertising personnel could be recruited to produce entertaining programs, especially for broadcast on television, radio and on the Internet as propaganda aimed at those who incite unrest during crises.

Chuck de Caro, a television producer, Air Force Academy graduate, former Green Beret and former CNN war correspondent, lectures about the power of television and his doctrine of warfare called Soft War. In 1997, NATO's Office of the High Representative (OHR) invited de Caro to Bosnia. He said the OHR/SFOR effort was uncoordinated, incoherent, crossed cultural and national lines, and had a mix of military, political and media elements that were trying to fight a tightly controlled, well-organized Serb foe with a focused propaganda message. Worse, CNN broadcast every NATO move to the Serbs in advance. De Caro said, "A cabal of fifth rate Balkan bozos... thumb their noses at the US and NATO" (Adams 1998: 286–7). De Caro suggested ignoring Serb leader, Karadzic and speaking directly to his people. Serbian Radio and Television (SRT), controlled by Karadzic, was well below the standards expected by US and European viewers. "Just above junior college campus TV level," de Caro said. The SRT showed clips of Nazi invaders during the Second World War, for example, intercut with pictures of SFOR troops, with the commentator calling them "SSFOR." It was crude but effective. De Caro argued that the United States should not counter-propagandize. It should counter-program by broadcasting the latest US programs with local messages that offer images of peace, stability and prosperity. "It's a ratings game," de Caro declared, "and I would win it hands down" (Adams 1998: 288).

Up until now, the US military has tried to control the flow of information instead of using it to aid US missions. A video feed, de Caro argued, that accurately portrayed the US government's position during a crisis and was made available free to every television station and newspaper over the Internet probably would be far more effective than any press briefing. Furthermore, there is no need actually to alter the images. The video would focus on certain things, not others, such as the benefits of cooperating with the United States and costs of an adversary's victory, just as the United States stressed during the 1948 Italian elections that dealt the Communists a severe setback.

The United States has begun to follow de Caro's advice, at least somewhat. In Somalia in 1993, the United States countered propaganda on Aideed's radio station with leaflets, a newspaper and a radio station that spread accurate news.

Upset that the new sources of information contradicted his station's accounts of events, Aideed agreed to tone down his propaganda. Conversely, trying to shut down Aideed's radio station, which the UN attempted to do later, was far more difficult than just countering his propaganda with accurate news from other sources (Clancy *et al.* 2004: 267–8).

In 2004, Washington established *Al Hurra*, the Free One, a satellite channel for Arab viewers in the Middle East. The station joined Radio *Sawa*, a US funded Arab radio station and *Hi*, an Arabic-English magazine, all designed to improve the US image in the Middle East (Associated Press 2004a: A13). Unfortunately, *Al Hurra* appears only to attract a tiny audience, perhaps 1 percent of the potential market. With more than 200 satellite television stations available, Arabs are not seeking more news outlets. The United States should not follow a Cold War model of the Voice of America, which provided news to the media-starved people of Eastern Europe but should instead produce quality programs with implicit pro-Western themes to attract an audience and provide the program to existing Arab television stations free or at a nominal cost (Lieven and Chambers 2006: B13).

The United States also used the marketing firm of McCann-Erickson to develop a $15 million propaganda program for the Middle East called "Shared Values" that was shown in Pakistan, Kuwait, Malaysia and on Pan-Arab stations in 2004. The spots featured documentary style interviews with Muslims in America who described the United States as tolerant. The State Department also distributed 5,000 copies of a book called *Muslim Life in America*. Unfortunately, the spots were seen as heavy-handed and, after opposition from the governments of Lebanon, Egypt and Jordan, were suspended (Murphy 2003: Online). The plan was good in theory, but an Arab public relations firm should have been used to create the spots, just as China, Kuwait and other nations hire US firms, who know the US public, to create propaganda for them to use in the United States. The United States also could use its vast foreign-born and educated population to create more effective propaganda for foreign audiences.

Propaganda messages need not be explicit and, in fact, must not be. Merely implicitly showing the benefits of democracy, a market economy and cooperation with the United States while including negative aspects of the United States and its policies to increase credibility would go far to weaning foreign publics away from local propaganda. Because local flavor is crucial, pro-American and, even more effectively, politically neutral local people should be recruited to produce spots that they believe will calm local conflicts or improve relations with the United States. Successful propaganda also must be entertaining so that target audiences choose to watch it. In a way, the United States is already winning the war for attention around the world as demonstrated by the dominance of its popular culture. The United States should use this powerful advantage to subtly portray the benefits of peace and cooperation with the United States.

As Crossman warned, however, psyops can accomplish little, if anything, in isolation. Such operations must be supported by appropriate diplomatic, economic and military policy that pay heed to the propaganda effects of such policies. Therefore, psyops officers should be involved in the development of all diplomatic, economic and military policies to ensure an effective psychological facet to such policies. If they are not involved, any US investment in psyops will be wasted.

The United States has used deception against adversaries in the past and should continue to do so, especially by taking advantage of the new possibilities offered by altered images. In terms of personnel, Machiavellianism tests can be used to determine which individuals have the potential to be effective deception officers, while individuals who were educated, have conducted business for long periods or who have traveled extensively in a target state can provide the much-needed empathy with a target population to increase the effectiveness of deception operations. In a similar vein, the United States was wise to use Iraqi–Americans to play the role of Iraqis during training at Fort Irwin, CA, for military units before they deployed to Iraq (Sahagun 2004: B1).

In the 1998 motion picture, *Wag the Dog*, top presidential aides conspire with a Hollywood producer to televise a mock war between the United States and Albania to distract attention from a growing presidential scandal. Compared with how altered images would probably be used, such an operation would be far too easy to disprove and would involve far too many people to ensure security. As with any conspiracy, the more people involved, the more chance for the truth to become known to the target. The opportunities to use altered images effectively to deceive or as propaganda, however, are vast, with a broad range of possible uses that carry little, if any, political costs.

Altered images can be used to make important targets appear damaged. In May 1998 a building along the 405 freeway in Los Angeles was draped with a huge mural that made it appear as if a massive hole had been blown in it. Traffic slowed as people gawked at what appeared to be a bomb-damaged building. The motion picture company that had created the effect for a shoot was ordered to remove the distracting image. The case, however, demonstrates how modern imagery could be used to make targets appear damaged, to camouflage potential targets or to create decoys, just as the British did with more rudimentary technology during the Blitz. In another twist, cardboard cutouts of images of guards, barriers and other defenses could be used to give the impression that a target was better guarded than it was, or to provide decoy targets to distract would-be attackers. Terrorists use photographs and video to help plan attacks (Zagorin and Shannon 2004: 5), creating an opportunity for US agents to alter images of targets by deleting certain defenses that could then be fed to terrorists to introduce errors into their planning.

Given that the United States is a land of immigrants, it is extremely well suited to using Americans as mimics for almost any other ethnic, religious or national group. Altered images could be used to support mimicry in many ways. An agent attempting to infiltrate an anti-American group could provide the terrorists with images of a high-value target to reinforce his credibility. If the images did not show defenses that were added just after the agent's photographs were taken, they would not jeopardize the agent's credibility if the terrorists learned of the new defenses. Altered images also could be used to support any bogus story a US agent was attempting to use to establish his bona fides with foreign agents or terrorists. For example, an agent mimicking an arms dealer and seeking to sell nuclear weapons could provide images of such weapons to interested parties as if the weapons were in his possession. Such images would be more powerful if an agent appeared in the images with the weapons or if the agent was able to produce a key part from the weapons to support his story.

Even given credibility concerns about lying to the media, the government could use the media in a deception operation involving mimicry. One platoon can mimic a division. Showing on CNN, for example, a platoon of the 101st Airborne Division at Fort Bragg, NC, while the rest of the unit was on its way to invade another state would contribute significantly to achieving surprise. In a similar operation, during the early stages of Operation Desert Shield, news reports and video of the arrival of a few US fighter aircraft in Saudi Arabia were presented as the arrival of entire, operational squadrons. It was weeks, however, before the jets' necessary support arrived. The few aircraft mimicked entire, well-supported units.

In one sense, dazzle is built into the US government's open and multi-faceted structure. Determining which information is important and which spokesperson is really stating what the government will do is a great challenge. Even for politicians or experts on the US government it is difficult to predict what policy the United States will pursue amid the plethora of options that are publicly discussed.

With CNN and an omnipresent media, operational security appears to be a thing of the past for the US military. The low point was reached early on 9 December 1993, when six Cobra gunships from US Navy ships took off to support a landing of US Navy SEALS and Marines in Somalia. As the helicopters approached the beach in the dark, they saw lights and flashes. At least one pilot armed his weapons system before an urgent message arrived; the flashes and lights were from press cameras and television crews. A top-secret mission was broadcast on CNN in real time (Adams 1998: 63). This type of coverage, however, can also be used to deceive. If a military operation was planned and the media, and therefore the target, knew it was imminent, the military could focus the media's attention on different types of attacks to dazzle the adversary. Information on amphibious, airborne and land-based attacks, different possible locations for an attack, as well as numerous timetables justified by credible factors, such as tides, phases of the moon, religious holidays and force readiness, could dazzle an adversary with so many possible attack options that they would have no idea which was the actual plan. Patton used dazzle during the Battle of the Bulge and there is no reason why such a stratagem would not work today. Furthermore, such an operation would not have the disadvantage of actually lying to the press, merely diffusing the media spotlight by directing it in several directions at once over the brief period before a military operation.

The possibilities for disinformation operations are broad, limited only by the creativity of deception officers. Altered images could be used to support almost any of the disinformation campaigns discussed in Chapter 7. For example, disinformation campaigns based on altered images could discredit a foreign politician, such as a devout Islamic leader who is rumored to be a playboy shown in an altered video drinking alcohol and carousing with scantily clad, unveiled women. Such images could be shown just before a military intervention, so there would not be time to discredit the images. To sow discord in another nation, the United States could clandestinely (black propaganda) distribute altered images of a paranoid and security conscious dictator's top aides or generals meeting with domestic rivals of the leader. A purge might result, just like the one the Gestapo helped trigger by feeding Stalin's paranoia about a possible military coup.

In 2003 the United States conducted an unplanned conditioning operation against Iraq. Before the war, the media focused attention on what was called a "shock and awe" campaign: a massive air attack to be launched at the start of the war. The irony was that the description was far better suited to the opening of the Gulf War than to the 2003 invasion. The later invasion began with a limited air attack that sought to assassinate Hussein. Then the ground attack began, in part to ensure control of the oil fields in the south, which the United States feared the Iraqis might set on fire. The air campaign that followed was limited. The Iraqis, however, appear to have been conditioned by all the media reports about "shock and awe" to expect a massive air attack preceding the ground assault, which did not happen. Such conditioning could be used in the future to deceive adversaries about the type, location and timing of an attack by focusing media attention away from the real plan.

Another case of unintended conditioning with far less favorable effects for the United States occurred in Iraq in 2003. After the Gulf War, the Shiites in southern Iraq rebelled against Hussein, but the United States did nothing to aid them and the dictator crushed the rebellion. Therefore, when the United States and Britain invaded Iraq in 2003 on the assumption that the people would welcome them, few did, especially not the Shiites in the south. The Shiites did not want to risk appearing to side with the United States or Britain, who might soon leave, exposing them to the wrath of Hussein or his successors. The United States and Britain attempted to counter this fear by distributing leaflets that said, "This time we won't abandon you," but their effect was minimal against the powerful effect of conditioning.

Also in an unintended way, the United States conditioned terrorists prior to 11 September 2001. After various terrorist attacks against the United States, including the 23 October 1983 truck bombing of the Marine barracks (and the French base) in Beirut, the 26 February 1993 bombing of the World Trade Center, the 7 August 1998 bombings of the US embassies in Kenya and Tanzania, and the 12 October 2000 attack on the USS *Cole*, the United States did little publicly in response, except for a handful of arrests and the installation of further defensive measures at embassies, ports and bases. The terrorists were probably conditioned to believe that the United States would do little militarily in response to such attacks. It was only after 11 September 2001 that the United States launched a full-scale military response in Afghanistan against the Taliban and Al Qaeda.

In the future, the United States should consider how it is conditioning its adversaries by the policies it pursues. Having a team of analysts role-play rival nations or groups to determine how US policy looks to them over time is one way to determine the effect of conditioning on such relationships, especially if the team members have extensive experience studying, living or working in the target state. For example, relations with Iran, Libya and North Korea, if viewed from a conditioning perspective, might lead to significant changes in US policy, since it sometimes appears that threatening behavior often results in a more profitable relationship with the United States than cooperative behavior. Bellicose behavior often leads to talks and deals that give the US adversary some of what it wants, while quietly cooperating often leads to the other nation being ignored by the United States.

The US military is currently in the midst of what it describes as a Revolution in Military Affairs (RMA). Much of this revolution is focused on network-centric warfare and networking weapons, intelligence and communications to increase combat effectiveness. The role of altered images in such a structure should not be ignored. Sapping an enemy's will to fight through propaganda and psyops, and misleading the enemy with deception operations should be part of the revolution, especially if altered images can be used in new ways. If altered images are included as a possible weapon in training, creative officers and troops will devise new ways to use them, some of which may prove effective. High Machs would be the most effective since they are unconventional thinkers who would be excellent at devising new tactics. Unfortunately, such thinkers are probably the least likely to fit into the peacetime military. A place, however, should be made for such eccentrics, even given the military's need for discipline and conformity. The British military and US Marines in the 1920s and 1930s should be studied to learn how they created a haven for eccentric thinkers, many of whom made significant contributions in the areas of psyops and deception operations, as well as to the development of amphibious and aircraft carrier tactics that helped win the Second World War.

The United States could also use all five types of altered images in the areas of diplomacy and warfare. The US military has often focused the media's attention on certain aspects of US operations to distract attention from other aspects of US operations. During the Gulf War, for example, the focus on a possible amphibious landing helped divert Iraqi attention from the Hail Mary left hook through the western desert. In a similar operation in 1982, the British in the Falklands used the media to confuse the Argentines about whether they would rely on a continued blockade, air attack or an invasion to end the war (Thompson 1985: 54–78). Such operations in the future could rely on the use of selective images, even as they avoided outright lies to the media.

The military also could provide far more access and, hopefully, coverage of civil works projects in nations it has intervened in, while downplaying any conflicts or combat. Providing professionally produced video free of charge would help increase the use of such positive, peaceful stories in the media, especially in the Third World. Locals should be used to produce and distribute such information to mask its source (gray propaganda). Finding creative local people would be crucial in order to avoid filming images that look staged, such as US soldiers passing out candy to children or US medics treating the elderly and sick. Such a strategy would also lessen the media's natural tendency to focus on combat and conflict during a military intervention, while also paying the additional dividend of learning what local people value. Such information then could be used to refine future propaganda.

Outright staging of events for propaganda purposes would harm US credibility, although events could be staged for deception operations in the intelligence realm. Staging, however, even for propaganda is a question of degree. The videotaping of the rescue of POW Private Lynch during the 2003 intervention in Iraq, parts of which were released to the media, were valuable in keeping the feel-good story in the public eye. Although certainly not staged, the videotaping was, of course, planned.

Other events, such as inviting Arab political and religious leaders to speak to Congress or in US houses of worship, for example, would go far to provide positive images about the United States in the Arab world and, especially, on such stations as Al Jazeera, the Qatar-based satellite television news station. Such events would be staged, yet they would be real and credible nonetheless, especially if religious, political or community leaders arranged the events instead of the US government.

Altering images or adding false captions would harm US credibility and should not be done for propaganda, although both forms of deception could be used in intelligence operations. Altered images or images with fake captions could be used, for example, to verify the authenticity of an agent provocateur attempting to infiltrate an anti-American group. Such images could "prove," for example, that the agent had killed US soldiers in Iraq, had access to certain types of weapons to sell or was a member of a large terrorist cell.

Altered images also could be used to build the case for military intervention. In a Machiavellian operation, images of some of the components for weapons of mass destruction could be mislabeled as belonging to a nation the United States was building a coalition against. Such images could be shown only to foreign leaders, thereby decreasing the chance that the deception would be exposed. If the public learned that the images had been altered, however, the US administration could blame the CIA, NSA or even a foreign government for providing the images to the White House, just as President Bush did, with apparently little political cost, with the inaccurate intelligence provided about WMDs before the war in Iraq in 2003.

The United States has extensive experience using photomontages in domestic political campaigns and should use them to support foreign policy goals and during military interventions. Video clips of life in North Korea intercut with images of life in South Korea would highlight the North's disastrous political and economic system. Images of Islamic terrorists blowing up bazaars and mosques could be intercut with images of US soldiers working with Iraqis to open bazaars and rebuild mosques to undermine support for the terrorists. Juxtaposing actual statements by an adversary, such as extreme Islamic fundamentalists and their beliefs about women not voting or attending school, and brutal legal codes combined with images of real Arab men and women suffering under such strictures could increase opposition to such groups. Such spots would be more credible if local people produced them about local issues.

Besides using altered images to achieve foreign policy and military goals, the United States also must be prepared to defend against their use by others. Members of the US military, intelligence services, State Department and even top politicians should be trained to understand the principles of deception, including the use of altered images. Such training could draw on the creative talents of wargamers, science fiction authors, script writers and video game creators. It should emphasize that truthful information can be used to support an invalid conclusion, and that an unreliable source may provide valid information for years before providing one crucial piece of information that points to an invalid conclusion. Unfortunately, training in deception does not translate into a strong defense against being deceived. Even so, leaders and analysts should be aware of how deception works, and that deception is relatively common and is extremely effective. The same principle applies to propaganda. Leaders must be aware that propaganda is extremely common and the many ways it can influence perceptions, beliefs and policy making.

Besides training in deception to at least be aware that it is used, gathering human intelligence (HUMINT) is the most effective method of exposing deception by uncovering the underlying plan. Deception resides in the minds of the adversary; therefore information about individuals is crucial to detecting deception. Satellite images will never show what is in the mind of an enemy. The US focus on high technology sensors does little good in detecting deception, since the same information can be valid or deceptive. Worse, more information is not necessarily better information. In order to detect deception operations, improving analysis and focusing on HUMINT is far more important than increasing the flow of information in general (Gerwehr and Glenn 2000: 57). For example, early in the Irish Troubles in 1969–71, the IRA was able to deceive the British with deceptive stratagems including disguises, diversionary attacks and disinformation. The British, however, gathered HUMINT and soon "patrol commanders were able to put a name to most of the faces they passed on the street" (Gerwehr and Glenn 2000: 53–4). With better maps and information about almost everyone they met, the British became far harder to deceive.

Counter-deception operations seek to use guile to uncover an adversary's deception operations. As Shakespeare counseled in *Hamlet*, "By indirections find directions out" (I, II, 66). The British in Northern Ireland, for example, used counter-deception operations. Special Air Service units in plain clothes went out in nondescript vans to gather information from local people (Gerwehr and Glenn 2000: 54). Government agents can also mimic rebels or terrorists to infiltrate enemy groups to learn about their deception plans and other operations from the inside. Also, if an infiltration operation is exposed, it harms the guerrillas or terrorists by weakening trust between them, since they will then wonder if any of their other comrades are government agents. Altered images could be used in such operations to provide proof that an agent provocateur is a real terrorist. Images of attacks on Americans could be used, for example, by Arab-American agents attempting to infiltrate Al Qaeda cells, just as Zig-Zag used camouflage and debris to fool German photoreconnaissance into believing that he had blown up a British aircraft works in the Second World War. In fact, altered images could be used to support the creation of an entire notional terrorist cell to attempt to infiltrate and obtain information about real cells.

The US preference to fight in alliances or through the United Nations makes it difficult to collect HUMINT, let alone conduct counter-deception operations. For example, in 1992 in Bosnia, the United Nations forbade the United Nations Protection Force (UNPROFOR) from gathering intelligence. "Intelligence is a dirty word in the United Nations," Air Marshal John Walker, the UK's Chief of Defense Intelligence, said. "The UN is not a thing in itself, it's an amalgam of 183 sovereign nations. If it does intelligence, it will be doing it against a sovereign UN member, so it's incompatible. But you need a military intelligence job to protect your troops. If you don't, you pay for it in body bags" (Adams 1998: 86). Attempting to run counter-deception operations in such an environment is next to impossible given the requirements for coordination and security between multiple states that often do not fully trust each other.

Although researchers are already developing programs to scan digital images for irregularities that can reveal they have been altered, the capability to alter images will probably always be one step ahead of such programs (Knapp 2004: Online). Even if such programs become good enough to detect any manipulated image, individuals will still be misled by mislabeled or staged images, cases of focusing on certain things and not others, and often will be affected by creative photomontages. Therefore, a crucial factor in defending against the use of altered images, whether for propaganda or to support deception operations, is to teach image literacy.

As Kraig Short (Personal Communication: 6 September 2005) argued, teaching image literacy and an ability to distinguish reality from virtual realities created by deceptive images goes against much of what Americans believe is "modern" behavior. Americans are taught that it is good to "get away from it all" and accept, if only for a little while, that what they are seeing in theaters or on their television or computer screens is "real." Usually such virtual realities are just entertainment. The problem is that those same virtual realities can be used to mislead the public on issues of national importance. For this reason, image literacy is crucial to future US foreign policy success and, more importantly, to the future of American democracy. The public must be able to distinguish truth from fiction and fact from lies or they will be at the mercy of knaves and, possibly, tyrants who use altered images to distort the presentation of reality for their own benefit.

Many states, such as China, Burma, Guatemala, Kuwait, Kazakhstan and Cameroon have hired public relations firms or prominent Americans to promote their policies and present a favorable image to the US public and government. The public should be aware of such strategies in order to understand how other states are attempting to influence American views of the world. Furthermore, given the rise of ethnic conflicts driven by propaganda, such as in Bosnia and Rwanda, and the rise of Islamic fundamentalism, the United States should focus on analyzing propaganda, especially propaganda using altered images, just as it did in the 1930s when it established a committee to monitor foreign media and expose Nazi propaganda. Today such a group could help educate the public about the use and effects of altered images when they are used with the intent to propagandize or deceive.

Image literacy should teach students about the power of images, and how images are emotion-based and not proposition-based. It should be stressed that images cannot prove something in the way words can prove a proposition. The influence of different typefaces, fonts, colors and visual styles, as well as the effect of accompanying music should be discussed so students understand their effects. Schools should teach students the standards different types of media have for the use of altered images so that they can better judge the validity of images. Image literacy education should also include teaching ways to detect altered images and that the source of an image should be considered, especially if the source has an agenda. Unfortunately, although there are textbooks on image literacy, such as Lester (2003), there is no manual for the lay person to learn to identify altered images. Most of the literature on altered images is concerned with how physically to manipulate images, not on how to detect actual alterations, let alone such stratagems as staging events, focusing on certain things and not others, changing captions, and the effects of photomontages.

It is far from easy to detect that a well-crafted altered image has been manipulated. Even experts often disagree about whether an image has been altered and the process of verifying an image can take months or even years. In 1932, for example, Gladys Cockburn-Lange, widow of a First World War pilot, published *Death in the Air: The War Diary and Photographs of a Flying Corps Pilot*. The book raised suspicions about its validity because no author was identified, no link was made between the pilot and the author, and there were no unit designations, last names, place names, specific dates or any information about the camera used to take the photographs in the book. Furthermore, it was extremely difficult to take photographs from planes during the First World War because early aircraft vibrated so much. Film speeds were also extremely slow, so it was impossible to take action shots without blurring, and cameras of the time also lacked the depth of field to photograph an entire dogfight. The images in the book, however, were clear shots from the air of aerial dogfights. Although doubt was cast on the veracity of the diary when it was published, nothing was proven until 1984, when the memorabilia of Wesley Archer and his wife were presented to the Smithsonian Air and Space Museum. Included in the materials were photographs of Archer working on aircraft models, which appear in the book, in a specially equipped photographic studio. Archer's wife had masqueraded as Gladys Cockburn-Lange (Brugioni 1999: 101–2; Nickell 1994: 62–4). It took 52 years to confirm that the diary was a hoax. In a similar case, a 1934 photograph of the Loch Ness monster published in a London newspaper was not disproved for 60 years (Brugioni 1999: 63).

In the early 1950s, a photograph of a Czech factory showed an assembly line of engines with 14 straight-through combustion chambers, like the Allison jet engine of the J-33 series made in the United States at that time. If the photograph was valid, then the Czechs and probably the Soviets had an engine to match the latest US model. The image raised fears in the United States that the Allison engine's plans had been stolen. US intelligence analysts looked through thousands of images and finally found a duplicate of the Czech photograph: an Allison jet engine plant in the United States. The Czechs had simply added coveralls to one worker, removed a tie from another worker, and painted a "No Smoking" sign in Czech on a pillar. The ruse was only detected, however, after thousands of hours of research (Brugioni 1999: 67). Such old cases based on rudimentary image-altering technology only serve to highlight how hard it is to detect images that have been manipulated digitally with today's high-powered computers.

It can take an enormous amount of time and money to determine whether a photograph is valid. Photo-interpreters, photogrammetrists (trained to obtain reliable information from images), chemical and paper analysts, experts on the subject of the image, and various technologists all can analyze images. Analysts look to see whether an image has had objects added or deleted, or if there has been any cropping or stretching, such as when models are made to look taller and more slender, as well as many other manipulation techniques. Often analysts use microscopes to search for any spurious lines or changes in a photograph or, even better, in a negative, since every duplication of an image decreases detail and makes it more difficult to detect that an image has been altered. If an analyst can see the entire roll of film that contained a suspicious image, it can sometimes reveal "practice" shots or missing frames where such shots were removed. Even if no such shots exist, images before and after the suspicious frame can sometimes shed light on the image under investigation. Today, with digital photography, however, there is often no negative to examine, making it even harder to determine whether an image has been altered.

There are certain factors that experts look for in a picture that laymen generally would not be able to notice, some factors they can at least be aware of, and some that they can use to attempt to detect altered images. Even being aware of the specialized methods used by photo-interpreters may, however, at a minimum, make the public aware of the methods experts use to detect fakes. Knowledge of such methods may make the public less likely to accept images and video at face value and more accepting of experts when they explain how they determine whether an image has been altered (Nickell 1994: 139–43).

One method of deception is to use a double to mimic another individual. Famous people throughout history have used stand-ins for various reasons.

George Washington, Hitler, Churchill, Stalin, Roosevelt, Montgomery, Idi Amin and Saddam Hussein all had doubles. The most effective method to unmask a double was devised by French statistician and criminologist, Alphonse Bertillon, between 1878 and 1880. The Bertillon system uses anthropometric measurements, such as eye color, hair, scars and deformities to determine if a photograph is actually of the person in question. Police still use the system today. In fact, other than a fingerprint, the ear is best for identification because no two ears are the same (Brugioni 1999: 152–4). On 19 December 1951, for example, North Korea released a photograph of a man purported to be Major General William F. Dean, who had been captured on 21 July. There were serious doubts about the photograph's authenticity. Although the man in the image appeared emaciated and gray compared with previous pictures of Dean, Arthur C. Lundahl, the dean of American photogrammetrists, after extensive analysis, said it was Dean. It turned out later that it was Dean in the picture, half starved from his captivity (Brugioni 1999: 85). The Bertillon technique, however, is far beyond what a layperson could do, although people should be aware that doubles are used and that there are methods to unmask them.

Experts may undertake a physical search to analyze a suspicious image, which involves going to where the photograph was allegedly taken to help determine if it was altered. For example, in 1991 an image taken in Laos of what was said to be an American Vietnam War POW appeared. To convince a family who believed their missing son was in the photograph, analysts went to Laos to find the man in the photograph. Unfortunately for the family, he turned out to be a 77-year-old Lao highland tribesman (Brugioni 1999: 94–5; Nickell 1994: 59–60).

Research can also determine the date on which an image was taken, which sometimes contradicts when a caption states the picture was taken. After the 1962 Cuban missile crisis, for example, some Americans feared that the Cubans were hiding missiles in caves. The CIA, however, had counted all 42 medium range ballistic missiles that the Soviets had brought to Cuba leaving on freighters. Even so, every known large Cuban cave was checked by aerial photography for any activity, such as truck tracks or recent construction. No such activity was found. A published photograph of Fidel Castro that showed him supposedly inspecting missiles in a cave was later found to be an image of him taken just after he took power in 1959 and was from a brochure promoting caves in Cuba for speleologists and tourists (Brugioni 109–10).

Experts can use film, photographic paper and even the chemicals used to develop an image to date a photograph (Brugioni 1999: 125). Such techniques are not possible in the age of digital images, but computer files often carry time and date information that can prove crucial to analyzing an image. Date–time–shadow analysis can also be performed in some cases to determine, if the exact location of the photograph is known, the exact date and hour an image was taken. Conversely, if the exact time is known, then the location where the image was taken can be determined. To perform this analysis, the analyst must be able to determine true north in the image, have a visible horizon, and shadows that begin and end in the frame. The image also must have been shot outdoors. This method was used to analyze photographs that proved that Commander Robert E. Peary was indeed at the North Pole "within the limits of his instruments" on 6 April 1909, when he said he reached the pole (Brugioni 1999: 86–7; Nickell 1994: 58–9).

The provenance or origin of an image can sometimes reveal clues that it is a fake. Unfortunately, US intelligence services tend to focus on the images themselves, and not on their provenance, which can be crucial (Epstein 1989: 233–4). Although such analysis often requires extensive research, individuals can consider the motives of the source and can look for logical inconsistencies that may suggest that an image has been altered. During the Cold War, for example, a Cuban publication showed a photograph of a US soldier standing guard at the White House to illustrate an article contrasting how violent Washington was to the tranquillity of Moscow. The soldier, however, wore a First World War-style helmet and carried an old rifle. The photograph had probably been taken at the start of the Second World War when First World War equipment was still being used (Brugioni 1999: 109–10). In a similar case, after the Soviets shot down a U-2 in 1960, they released images of the aircraft that showed rows of rivets. Kelly Johnson, the U-2's designer, however, pointed out that the U-2 was flush-riveted. The images could not have been of a U-2, although it remains unclear why the Soviets released images of a different plane when they already had the U-2 wreckage (Brugioni 1999: 130–2). More recently, when the Turks invaded Cyprus in 1974, Greek Cypriots charged the British with complicity in the invasion and published photographs they claimed supported their allegations. One photograph was of the destroyer HMS *Devonshire* supposedly outside the Cypriot port of Kyrenia. The ship in the photograph, however, carried the pendant number F57. "F" denoted frigate, however, not destroyer. Furthermore, F57 was HMS *Andromeda*, which it was relatively easy to prove was nowhere near Cyprus during the invasion (Brugioni 1999: 131–3).

Flora and fauna, as well as geography can help determine if an image or film is valid. In 1986, a film of the nuclear accident at Chernobyl appeared on several US television networks. The film showed a mountain range nearby, although Chernobyl is on a river plain and also showed a large city, when Chernobyl is, in fact, a small town. After some research, it was determined that the film was of Trieste, Italy. *20/20* and *Nightline*, both of whom had used the film, acknowledged the error (Brugioni 1999: 97).

Comparing a suspect image to ones of known veracity can sometimes reveal that an image has been altered. The Soviet Union, for example, often published altered images. Soviet editors, however, had some freedom with which photographs they published and how they were altered, so Western investigators sometimes determined that a photograph had been altered by comparing images of the same thing from different publications (Brugioni 1999: 63, 141). In today's media-saturated world, individuals may be able to compare images of the same thing from different media sources to help determine whether an image has been altered. For example, shoppers in supermarket checkout lines may have noticed during O.J. Simpson's murder trial that the ex-football player's mug shot on the cover of *Time* had been darkened compared with the same image on the cover of *Newsweek*. *Time* had altered the image to make it appear more dramatic.

The relative size of objects in an image can be compared with each other and the perspective or spatial relationship between objects analyzed. When poorly done, laymen can sometimes notice such discrepancies. The United States often unmasked Soviet photomontages by comparing the size of known objects in an image to other objects in the same image. Using this method, an image of MiG-15s flying over a Skoryi destroyer was proven to be a montage. The United States knew that the wingspan of the jets was 37 feet 6 inches, yet, at the same scale, the destroyer would be 161 feet. The United States, however, knew that the destroyer was 420 feet long, proving that the image was a photomontage (Brugioni 1999: 72–3).

Laymen can look closely at shadows to analyze an image that was taken outdoors. Shadows can be crucial, since they must all fall in the same direction and be consistent in size and shape with the object photographed. In the early 1950s, for example, a Chinese photograph of a People's Liberation Army-Navy Landing Ship Tank and landing craft was shown to be a montage, since the shadows on the landing craft in the image did not all fall in the same direction (Brugioni 1999: 71). Halation, the fog or halo in a photograph around the image of a highly reflective surface or light source, can also be used by experts and laymen to detect an altered image. Halation around one object but not around similar objects may suggest that an image has been altered.

Vanishing point analysis can be performed by laymen with a ruler. In an image, lines that are parallel should appear to converge in the distance. In an altered image, vanishing points are often inadvertently displaced. For example, if a building has been added to a set of buildings, their rooflines may not converge in the distance, as they should (Brugioni 1999: 89–90).

Crop lines and tones can also give the photo-manipulator away to experts and laymen. Trees are especially hard to alter since the manipulator either has to cut around all the leaves as a single object or cut around each individual leaf and limb, which is extremely difficult to do (Brugioni 1999: 74). Experts use microscopes to search for cut lines and to determine if there are changes in the grain suggesting that two images have been combined. Individuals can also look closely for evidence of cropping, although computers make such alterations extremely difficult to detect. In terms of tone, if photographs are combined, the backgrounds are often harsh or flat. Analysts and laymen can also look for the same object or person in a crowd appearing several times in the same image. During the Cold War, for example, the Soviets published a photograph of Soviet Mail seaplanes flying over a guided missile patrol boat squadron. It turned out that one photograph of four Mails had been copied and then half of the image was copied again to make it appear that 10 Mails were flying over the boats (Brugioni 1999: 111). Similarly, China, during Mao's Great Leap Forward, published a photograph showing a huge Chinese oil field at night to illustrate that China was decreasing its dependence on foreign oil. The photograph, however, showed the same set of lights repeated several times (Brugioni 1999: 93–4).

Laymen can also pay attention to the depth of field to analyze an image. Depth of field depends on the focal length of the camera lens, size of the camera aperture and distance of the camera from the object. Luckily for photo-interpreters and laymen, it is extremely difficult to get depth of field right when images are combined. Usually there are too many things in focus. If objects in the foreground and background are all in perfect focus, it is probably a fake. Furthermore, in color photographs, objects in the distance are generally bluer than objects in the foreground, which is difficult to accomplish in an altered image (Brugioni 1999: 73–4).

Sometimes Newton's third law of physics – for every action there is an equal and opposite reaction – exposes a fake. A photograph of a Soviet Yak-24 Horse helicopter, for example, above a crowd at an air show was suspicious, since not a single person in the crowd was looking up at what must have been a loud helicopter directly overhead (Brugioni 1999: 81).

Close inspection of an image can lead to the discovery of inconsistencies that the manipulator of the image may have overlooked. If an object or person has been removed, the corresponding shadow may still be present or part of the object or person may remain. In some Soviet photographs, for example, people were removed, but their shoes remained (Brugioni 1999: 105). In another case, the Soviets were building a new bomber. They usually produced a civilian model of the bomber as an airliner, in this case the Tu-20 (later Tu-95) Bear long-range bomber and the commercial version, the Tu-114 Cleat. The Soviets published a photograph of a factory building the Cleat. It was meant to show one Cleat under construction, but the photo-manipulator failed to delete the tails of two more Cleats that showed above the first Cleat's fuselage (Brugioni 1999: 105).

Although it is difficult, the public, which already does spot some manipulated images in tabloids and on the Internet, can be trained to detect some altered images. Most such easily identified altered images, however, are poorly done; while even some professionally altered images are given away by their suspicious content, source or medium. An increasing number of altered images, however, are created by experts and are extremely difficult to detect. A questioning mind is the first line of defense against the use of altered images for propaganda and deception. The public should therefore develop a better understanding of how images can be altered, their effects, and how experts analyze images to determine whether they have been altered. If they do not, they will fall prey to those who use altered images for propaganda and to deceive.

CONCLUSION

The past is rife with examples of psyops, propaganda and deception operations that had significant effects on politics, diplomacy and the outcome of wars. Altered images played a part in a few of those cases, yet the future of psyops, propaganda and deception operations promises the use of far more images that have been manipulated. With computers providing the capability quickly and easily to alter images and video seamlessly and also add appropriate audio and create images of events that never occurred, altered images are set to become a crucial component of propaganda and deception operations. Past examples of psyops and deception operations provide just a hint of the vast range of possibilities for new, deceptive realities that manipulated images can now provide for those who create propaganda and seek to deceive. Altered images, whether still or moving, can be used for propaganda or to deceive by focusing on certain things, not others, staging a scene, altering an image, false captioning or by creating a photomontage. Furthermore, with the rise of a global communications network based on the spread of television and the Internet, altered images can now be disseminated globally almost instantly. The image revolution of the twentieth century combined with the relative lack of public skill in detecting manipulated images has ushered us into the age of altered images.

Although IW focuses on computers and their use in warfare, only a small part of the literature in the burgeoning field deals with perception management. It is in this area that altered images could play a major role, although the IW literature devotes little, if any, attention to their potential. This area demands more attention because altered images are already being used. Although the news media has strict rules against manipulating images, altered images have become common on the Internet, in motion pictures and advertisements, in tabloids and in some types of magazines, as well as, albeit less commonly, in the news media – at least, so far. Most importantly, the ease with which a false story supported by altered images can be disseminated has greatly increased due to the Internet and changes in the media. These changes make this the perfect time for the use of altered images in propaganda and to support deception operations, especially for use against the United States and other media-saturated societies.

Because of their technological lead in many of the fields related to image manipulation, the United States and the West will probably be leaders in the use of altered images for propaganda and to support deception operations. This advantage, however, does nothing to prevent others from using manipulated images against the United States and the West. The United States is especially vulnerable to deception or propaganda based on altered images because America is a media culture based on images. A.M. Willard's 1876 revolutionary painting, the "Spirit of '76," the flag raising on Iwo Jima, the Zapruder film of John F. Kennedy's assassination and images of Neil Armstrong taking man's first steps on the moon are some of the images that are the cultural landmarks of America. Musicians today are rarely popular unless they can not only play entertaining music, but also look handsome or beautiful in videos. Motion pictures are covered as if their stars, releases and grosses are the most important things on earth. Scenes from motion pictures are the touchstones of American society, such as Cary Grant being buzzed by a biplane in *North by Northwest*; John Wayne framed in a cabin doorway in *The Searchers*; Marilyn Monroe standing on a subway grate as the wind blows her dress up in *The Seven Year Itch*; and the battlestar exploding in *Star Wars*.

Furthermore, America's leaders, showing the hubris common to all great powers, believe that they cannot be deceived, which instead makes them the perfect marks for deception. Worse, with far more capability to collect intelligence information than any other state, the United States has countless avenues along which an adversary can pass disinformation based on altered images to US decision-makers. The all-encompassing media also provides a feedback mechanism to judge how a bogus story is playing in the United States.

Such deception-based attacks have already occurred. The KGB, for example, was successful at starting bogus stories linking the United States to the creation of HIV/AIDS, using Latin American babies for organ transplants and employing a Korean civilian airliner as a reconnaissance aircraft over the Kamchatcka peninsula. A British newspaper has published altered images of British troops allegedly abusing Iraqi detainees, while John Kerry appeared in an altered image seated near Jane Fonda at an anti-Vietnam War rally. Such altered images will only become more common as the capability to manipulate images improves and spreads.

The United States, a nation of engineers, builders and businesspeople, pays scant attention to issues of propaganda, let alone of deception, and largely ignores their effect on the US public, let alone their effect on allies, neutral countries and adversaries. The future threat to the United States in this area, however, is only limited by the creativity of America's adversaries. Many of those who devise operations to use altered images against the United States will probably have been educated at US universities or have worked in the United States, thus ensuring that they have the empathy necessary to create effective propaganda and deception operations targeted at Americans. Given that propaganda and deception operations are time and culture dependent, it is impossible to predict specific future operations, but such operations will follow the same principles as those of the past, which have been extremely effective. It is a certainty that such operations will be used against the United States in the future, since the weak have always used deception and propaganda against the strong and they now have the added weaponry of computer-aided manipulation of images and audio, and a global distribution network based on the Internet and the global media.

Legal, ethical and political constraints limit the use of altered images by the United States. These limitations, however, should not bar the United States from using certain kinds of altered images in certain situations. The crucial variables are the degree to which the US public is involved, the type of altered image used and the perceived graveness of the threat. As a general rule, if there is a chance that the public will see them, the United States should not use false captions or actually manipulate images, especially in propaganda. Because the public is less likely to see the altered images used in certain types of deception operations, especially those targeted at foreign leaders or militaries, however, the use of altered images for these purposes should have far more leeway than their use for propaganda. The public does, however, appear to accept the government's use of certain types of altered images, whether for propaganda or deception operations. Photomontages and staged events are already commonly used in US domestic political campaigns and could be used to promote US interests overseas or to weaken support for foreign adversaries. The public also does not appear to object in most cases when the government focuses media attention on certain things while ignoring other things, especially in relation to military operations. Altered images also can and should be used to camouflage targets and to act as decoys for military and intelligence purposes because the public is rarely involved in such operations.

The gravity of the threat also plays a role in determining the leeway the US government has in the use of altered images. The United States and other democracies have lied to their own people in the past, both lies of omission and lies of commission, during times of war when the threat seemed high, and even when the stakes were not that high, such as during military interventions. Furthermore, such deceptions usually had little or no political cost, suggesting that the US public is willing to allow the government great latitude in the use of certain types of altered images during military interventions and wars, especially those related to military operations.

Given its expertise in the areas of television, motion pictures, advertising, radio, public relations and the Internet, the United States should counter-program against the propaganda generated by adversaries. By producing propaganda that is entertaining, audiences will tune in to US messages and tune out those of its foes. US popular culture already engulfs the rest of the world and there is no reason why, during a crisis, skillful US propaganda promoting peace and cooperation with promises of peace and prosperity could not overwhelm the propaganda of those who seek to incite disorder, conflict or genocide.

Relying on its vast pool of immigrants, Americans who have special language and area knowledge should be placed in key positions to implement psyops and deception operations. Some such individuals would be well suited to infiltrating anti-American groups that are planning deception operations or terrorist attacks against the United States. It was extremely unfortunate that the United States did not take full advantage of Iraqi-Americans and other Arab-Americans who offered their services during the 2003 conflict in Iraq. Iraqi-Americans probably would have been effective propagandists during and after the war in 2003, especially if they had worked with Iraqis to produce propaganda with local flavor. They could have provided insights into Iraq, both to help win the war and, more importantly, to help in the attempt to create a peaceful post-war Iraq.

A nation, especially if it seeks to be a great one, must stay the prudent, wise course. If the initial reasons for intervening or waging war were valid, a small number of casualties, even if they are extensively covered by the media, should not cause a nation to veer from its goals as the United States did in Somalia. When the body of a single American was dragged through the streets of Mogadishu, the US government changed policy and withdrew. A great nation deserves – needs – great leaders who, after carefully and prudently deciding that military intervention is necessary, will stay the course, even if casualties are incurred. The bar should be placed high for leaders to decide to use military force, so that, once committed, lives are not wasted for goals that are not important enough to justify the loss of American lives.

A great nation also deserves leaders who are aware of the influence of deception and propaganda, especially today when altered images can so easily and often subtly distort decision-making. Deception training should be required for personnel in the military, CIA, NSA and the State Department, and probably for all politicians who must make decisions based on information that may be deceptive or propaganda. Unfortunately, another state's propaganda is often not even recognized as such. China and Saudi Arabia, for example, are just two of the many states that carefully craft their image in the United States without their messages being tinged with the negative connotation of being propaganda, yet strongly affect US decision-making and policies. Training of US leaders should seek to remedy this often-unnoticed influence on US foreign policy decision-making. Leaders must be aware of how altered images can deceive or they may, as the Bush administration appeared to do in 2003 in relation to Iraq, use military force and risk the lives of US military personnel based on erroneous information. Leaders must see the world clearly and accurately in order to pilot the ship of state prudently, wisely and safely. If they do not, disaster looms dead ahead.

Since in a democracy, the public has the final decision at the ballot box, image literacy should be included in the educational system to ensure that US citizens, as well as the media and government have the ability to distinguish truth from falsehood when information is presented, as it usually is today, in the form of images. At a minimum, the public, the media and US leaders should understand the power of images and how they influence beliefs without making testable propositions or arguments. Americans must understand the impact of images, while not granting them more influence than they deserve, especially when they present a distorted picture of reality. Education is crucial in this respect.

Unfortunately, there has been no basic change in the education system since the 1700s, when text was dominant and rhetoric, grammar and literacy were emphasized in schools. Text is no longer dominant; images are the form by which most people now receive most of their information. Therefore, both to counter the threat of enemies using altered images against the United States and to stay true to the tenets of democracy as envisioned by the Founding Fathers, image literacy should become a key feature of education in the United States. Such an education would help citizens and leaders identify altered images and better understand the influence, strengths and weaknesses of images themselves. Failure to change the US education system to include image-based literacy will betray the basic requirement for citizenship as outlined by the Founding Fathers, casting the future of the Republic in doubt: citizens must be able to rationally evaluate information in order to fulfill their role as voters. Today, images dominate our lives and mold our vision of reality. We must, therefore, be aware of how many of these images distort our view of the reality we think we see and affect decision-making, often adversely and sometimes with dire consequences.

With ever-improving capabilities to create new realities with altered images, video and sound, the future of propaganda, psyops and deception operations offers unlimited vistas for adversaries to deceive US leaders and to reinforce selected beliefs with propaganda. Conversely, such techniques also offer great potential for the United States to use psyops, propaganda and deception operations against its adversaries. The United States should use these methods even if it is militarily far stronger than its adversaries in order to lower the cost of military interventions and wars, often for both sides, as well as to help achieve foreign policy goals.

Today, the United States is the most powerful nation militarily. Its sword is by far the strongest and best funded of any wielded on earth. As Napoleon Bonaparte said, however, "There are but two powers in the world, the sword and the mind. In the long run the sword is always beaten by the mind." Unless the United States uses creative, agile and devious minds in the realm of psyops and deception operations to help wield its sword, rivals and enemies, who use creative, agile and devious minds to create psyops and deception operations based on altered images, may well be able to severely blunt, if not break, the American sword.

BIBLIOGRAPHY

Achenbach, J. (1998) "Reality check," in Alberts, D.S. and Papp, D.S. (eds) *The Information Age: An Anthology on its Impact and Consequences*, Washington, DC: National Defense University Press, Ch. 6.

Adams, J. (1998) *The Next World War: Computers are the Weapons and the Front Line is Everywhere*, New York: Simon and Schuster.

Aldrich, R.J. (2002) *The Hidden Hand: Britain, America, and Cold War Secret Intelligence*, Woodstock, NY: Overlook, 2002.

Amato, I. (July/August 2000) "Digital fraud in real time," *Technology Review: MIT's Magazine of Innovation*. Available at www.techreview.com/articles/july00/amato.htm (accessed 7 July 2005).

Andrew, C. and Gordievsky, O. (1990) *KGB: The Inside Story*, New York: Harper Collins.

Andrews, M. (1958) "Psychological warfare in the Mexican War" in Daugherty, W.E. and Janowitz, M. (eds) *A Psychological Warfare Casebook*, Baltimore: Johns Hopkins University Press, 72–3.

Arkin, W. (2004) "Treasure hunters need reliable maps," *Los Angeles Times*, 8 February, M1–M6.

Associated Press (2004a) "New US. Channel Airs on Arab TV," *Los Angeles Times*, 15 February, A13.

——— (2004b) "UK: abuse photos not taken in Iraq," Online. Available from http://www.chinadaily.com.cn/english/doc/2004–05/14/content_330682.htm (accessed 18 May).

——— (2004c) "Newspaper apologizes for 'fake' photos, editor steps down," Online. Available from http://www.signonsandiego.com/news/world/iraq/20040514-1037-britain-dailymirror.html (accessed 19 May 2004).

——— (2004d) "Americans object to war images online," Online. Available from http://www.boston.com/business/technology/articles/2004/07/11/americans_object_to_war_images_online/Netscape.com (accessed 18 July 2004).

Badam, R. (2005) "Bollywood in uproar over sex scandal," *Los Angeles Times*, 16 March, E11.

Baran, S. and Davis, D. (1999) *Mass Communication Theory: Foundations, Ferment and Future*, 2nd edn, Belmont, CA: Wadsworth Publishing.

Baron, R., Byrne, D. and Watson, G. (1995) *Exploring Social Psychology*, Scarborough, Ont.: Allyn & Bacon.

Becker, H. (1958) "Nature and consequences of Black propaganda" in Daugherty, W. and Janowitz, M. (eds) *A Psychological Warfare Casebook*, Baltimore: Johns Hopkins University Press, 672–8.

Bennett, B. and Weiskopf, M. (2004) "Inside the takedown," *Times*, 7 June, 36–7.

Berger, A. (1995) *Essentials in Mass Communication Theory*, Thousand Oaks, CA: Sage.

Berger, C. (1976a) "Selection of themes" in McLaurin, R., Rosenthal, C. and Shillings, S. (eds) *The Art and Science of Psychological Operations: Case Studies of Military Application*, vol. 2, Washington, DC: American Institutes for Research, 670–3.

——— (1976b) "The famous 'MIG' leaflet" in McLaurin, R., Rosenthal, C. and Shillings, S. (eds) *The Art and Science of Psychological Operations: Case Studies of Military Application*, vol. 2, Washington, DC: American Institutes for Research, 689.

Berreman, J. (1958) "Assumptions about America in Japanese propaganda to the United States" in Daugherty, W. and Janowitz, M. (eds) *A Psychological Warfare Casebook*, Baltimore: Johns Hopkins University Press, 430–41.

Blackstock, P. (1958) "German psychological warfare against the Soviet Union 1941–1945" in Daugherty, W. and Janowitz, M. (eds) *A Psychological Warfare Casebook*, Baltimore: Johns Hopkins University Press, 263–72.

Blainey, G. (1988) *The Causes of War*, 3rd edn, New York: The Free Press.

Boot, M. (2005) "Mubarak, $2 billion and change," *Los Angeles Times*, 10 February, B13.

Boucher, G. (2005) "Voices from beyond," *Los Angeles Times*, 14 April, A1, A20.

Bowden, M. (1999) *Black Hawk Down: A Story of Modern War*, Boston: Atlantic Monthly Press.

Bowdish, R. (1999) "Information-age psychological operations," *Military Review*, December–February, 28–36.

Bowyer, J. (1980) *Cheating*, New York: St. Martin's.

Brack, D. (1996) "Do photos lie?" *U.S. Naval Institute Proceedings*, August, 47–9.

Breuer, W. (1988) *The Secret War with Germany: Deception, Espionage and Dirty Tricks 1939–1945*, Novato, CA: Presidio.

Brooks, M. (2002) *Military History's Most Wanted*, Dulles, VA: Brassey's.

Brown, A. (1975) *Bodyguard of Lies*, New York: Harper and Row.

Brugioni, D. (1999) *Photo Fakery: The History and Techniques of Photographic Deception and Manipulation*, Dulles, VA: Brassey's.

Butterfield, L. (1958) "Psychological warfare in 1776: the Jefferson-Franklin plan to cause Hessian desertions" in Daugherty, W. and Janowitz, M. (eds) *A Psychological Warfare Casebook*, Baltimore: Johns Hopkins University Press, 62–72.

Camera/Iraq (2003) "London 'Evening Standard' runs a cover photo presumably showing Baghdad Iraqis cheering the downfall of Saddam Hussein," 9 April, Online, available at http://www.camerairaq.com/2003/06/london_evening_.html (accessed 3 March 2006).

Carroll, W. (1958) "Where is the Luftwaffe" in Daugherty, W. and Janowitz, M. (eds) *A Psychological Warfare Casebook*, Baltimore: Johns Hopkins University Press, 373–80.

Chen, E. (2004) "New strides in presidential stagecraft," *Los Angeles Times*, 6 July, 15.

Choukas, M. (1976) "Postal subversion" McLaurin, R., Rosenthal, C. and Shillings, S. (eds) *The Art and Science of Psychological Operations: Case Studies of Military Application*, vol. 2, Washington, DC: American Institutes for Research, 726–7.

Cialdini, R. (1995) "Principles and techniques of social influence" in Tesser, A., *Advanced Social Psychology*, New York: McGraw-Hill, 257–81.

Clancy, T., Zinni, T. and Koltz, T. (2004) *Battle Ready*, New York: Berkley Books.

Cobb, C. (2003) "Digital fakery: the visually 'perfect' body is but a click away – even for Titanic Kate and our Nelly," *The Ottawa Citizen*, 2 February, Online, available at http://www.reclaimyourculture.com/Assets/Resources/Beauty/Digital%20fakery_files/Digital%20fakery.doc (accessed 3 February 2003).

Cockburn, A. (1984) *The Threat: Inside the Soviet Military Machine*, New York: Random House.

——— (2000) "CNN and psyops," *Counterpunch*, 26 March, Online, available at http://www.counterpunch.org (accessed 28 August 2002).

Codevilla, A. (1989) "Political warfare" in Lord, C. and Barnett, F. (eds) *Political Warfare and Psychological Operations: Rethinking the U.S. Approach*, Washington, DC: National Defense University Press, 77–101.

Cook, G. (2002) "At MIT, they can put words in our mouths," *The Boston Globe*, 15 May, A1.

Cotler, G. (1958) "Kobayashi," in Daugherty, W. and Janowitz, M. (eds) *A Psychological Warfare Casebook*, Baltimore: Johns Hopkins University Press, 198–224.

Daniszewski, J. (2004) "N. Ireland power-sharing pact foiled by photograph," *Los Angeles Times*, 9 December, A20.

Dasbach, A. (1976) "US–Soviet magazine propaganda: America illustrated and USSR," in McLaurin, R., Rosenthal, C. and Shillings, S. (eds) *The Art and Science of Psychological Operations Case Studies of Military Application*, vol. 2, Washington, DC: American Institutes for Research, 720–2.

Daugherty, W. (1958a) "The creed of a modern propagandist" in Daugherty, W. and Janowitz, M. (eds) *A Psychological Warfare Casebook*, Baltimore: Johns Hopkins University Press, 35–46.

——— (1958b) "Benno Frank" in Daugherty, W. and Janowitz, M. (eds) *A Psychological Warfare Casebook*, Baltimore: Johns Hopkins University Press, 248–51.

——— (1958c) "Bomb warnings to friendly and enemy civilian targets" in Daugherty, W. and Janowitz, M. (eds) *A Psychological Warfare Casebook*, Baltimore: Johns Hopkins University Press, 359–62.

——— (1958d) "Commander Norden and the German admirals" in Daugherty, W. and Janowitz, M. (eds) *A Psychological Warfare Casebook*, Baltimore: Johns Hopkins University Press, 494–7.

——— (1958e) "Truth vs credibility" in Daugherty, W. and Janowitz, M. (eds) *A Psychological Warfare Casebook*, Baltimore: Johns Hopkins University Press, 611–14.

——— (1958f) "A divisive appeal to the CCF that was never made" in Daugherty, W. and Janowitz, M. (eds) *A Psychological Warfare Casebook*, Baltimore: Johns Hopkins University Press, 635–41.

——— (1958g) "An old technique employed in a modern setting," in Daugherty, W. and Janowitz, M. (eds) *A Psychological Warfare Casebook*, Baltimore: Johns Hopkins University Press, 643–4.

——— (1958h) "Credibility in leaflet and poster illustrations" in Daugherty, W. and Janowitz, M. (eds) *A Psychological Warfare Casebook*. Baltimore: Johns Hopkins University Press, 751–3.

Daugherty, W. and Janowitz, M. (eds) (1958) *A Psychological Warfare Casebook*, Baltimore: Johns Hopkins University Press.

Davies, F. (2003) "The power of propaganda," Knight Ridder Newspapers, 16 October, Online, available at http://www.utne.com/webwatch/2003_119/news/10905-1.html (accessed 3 February 2006).

Davis, M. (1976) "Visual communications in the Nigerian Civil War" in McLaurin, R., Rosenthal, C. and Shillings, S. (eds) *The Art and Science of Psychological Operations: Case Studies Of Military Application*, vol. 2, Washington, DC: American Institutes for Research, 698–9.

Deighton, L. (1980) *Blitzkrieg: From the Rise of Hitler to the Fall of Dunkirk*, New York: Knopf.

Demick, B. (2005) "Secret N. Korean footage suggests nascent dissent," *Los Angeles Times*, 23 May, A1, A6.

DiMassa, C. and Hollis, R. (2004) "Man fakes own beheading video," *Los Angeles Times*, 8 August, B4.

Dorril, S. (2000) *MI6: Inside the Covert World of Her Majesty's Secret Intelligence Service*, New York: The Free Press.

Drogin, B. (2004) "Suspicion of Chalabi deception intensifies," *Los Angeles Times*, 23 May, A1, A7.

Drogin, B. and Miller, G. (2005) " 'Curveball' debacle reignites CIA Feud," *Los Angeles Times*, 2 April, A1, A19.

Effross, W. (2000) "A dressing-down for digital drag: image manipulation on the internet," Washington College of Law website, Online, available at http://www.wcl.american.edu/faculty/effross/dustin.cfm (accessed 7 July 2004).

Epstein, E. (1989) Deception: the invisible war between the KGB and the CIA, New York: Simon and Schuster.

Fairness and Accuracy in Reporting (2004) "Reagan: media myth and reality," Online, available at http://www.fair.org/index.php?page = 1832 (accessed 7 February 2006).

Farago, L. (1958a) "Research and analysis: the chief components of usable intelligence" in Daugherty, W. and Janowitz, M. (eds) *A Psychological Warfare Casebook*, Baltimore: Johns Hopkins University Press, 514–16.

——— (1958b) "Black propaganda: its techniques and requirements" in Daugherty, W. and Janowitz, M. (eds) *A Psychological Warfare Casebook*, Baltimore: Johns Hopkins University Press, 670–2.

Foss, S. (2004) "Framing the study of visual rhetoric: toward a transformation of rhetorical theory" in Hill, C. and Helmers, M. (eds) *Defining Visual Rhetorics*, London: Lawrence Elbaum, 303–13.

Frank, R. (2003) "Altered photos break public's trust in media," *Los Angeles Times*, 7 April, B11.

Freelance (1996) "Standard *loses bottle*," *freelance*, Online, available at http://media.gn.apc.org/9605mani.html (accessed 3 February 2006).

Gelernter, D. (2005) "When pictures lie," *Los Angeles Times*, 9 September, B11.

Gerwehr, S. and Glenn, R. (2000) *The Art of Darkness: Deception and Urban Operations*, Santa Monica, CA: RAND.

——— (2003) *Unweaving the Web: Deception and Adaptation in Future Urban Operations*, Santa Monica, CA: RAND.

Gessner, P. (1976) "Films from the Vietcong" in McLaurin, R., Rosenthal, C. and Shillings, S. (eds) *The Art and Science of Psychological Operations: Case Studies of Military Application*, vol. 2, Washington, DC: American Institutes for Research, 680–2.

Glionna, J. (2004) "Vietnam War-era photo seen as a bid to tarnish Kerry," *Los Angeles Times*, 12 February, A24–5.

Goldberg, J. (1989) "Comment" in Lord, C. and Barnett, F. (eds) *Political Warfare and Psychological Operations: Rethinking the U.S. approach*, Washington, DC: National Defense University, 160–3.

Govier, T. (1988) "Are there two sides to every question?" in Govier, T. (ed.) *Selected Issues in Logic and Communication*, Belmont, CA: Wadsworth Publishing, 43–54.

Greene, B. (2004) "Space and time, strings attached," *Los Angeles Times*, 7 March, R4.

Hall, W. (2003) *Stray Voltage: War in the Information Age*, Annapolis, MD: Naval Institute Press.

Hallion, R. (1992) *Storm Over Iraq*, Washington, DC: Smithsonian Institution Press.

Hamburger, T., Anderson, N. and Miller, C. (2005) "Tax-funded White House PR effort questionable," *Los Angeles Times*, 8 January, A1, A14.

Hansen, J. (2002) "Soviet deception in the Cuban missile crisis" in *Studies in Intelligence*, 46/1, 49–58.

Hardwig, J. (1988) "Relying on experts" in Govier, T. (ed.) *Selected Issues in Logic and Communication*, Belmont, CA: Wadsworth Publishing, 125–37.

Harris, E. (1967) *The "Un-American" Weapon: Psychological Warfare*, New York: Lads.

——— (1976) "Operation strike" in McLaurin, R., Rosenthal, C. and Shillings, S. (eds) *The Art and Science of Psychological Operations: Case Studies of Military Application*, vol. 1, Washington, DC: American Institutes for Research, 412.

Henderson, I. and Goodhart, P. (1958) *Man Hunt in Kenya*, New York: Bantam.

Herz, M. (1976) "Brief observation on the importance of up to date language in 'black' or 'grey' propaganda leaflets" in McLaurin, R., Rosenthal, C. and Shillings, S. (eds) *The Art and Science of Psychological Operations: Case Studies of Military Application*, vol. 2, Washington, DC: American Institutes for Research, 678–9.

——— (1982) "Lessons from VC/NVA propaganda" in McLaurin, R. (ed.) *Military Propaganda, Psychological Warfare and Operations*, New York: Praeger, 351–5.

Hiebert, R. (1995) "Editor's note" in Hiebert, R. (ed.) *Impact of Mass Media on Current Issues*, 3rd edn, White Plains, NY: Longman, 335.

Hill, C. and Helmers, M. (2004) "Preface" in Hill, C. and Helmers, M. (eds) *Defining Visual Rhetorics*, London: Lawrence Elbaum, ix–xi.

Hinsley, F., Thomas, E., Ransom, C. and Knight, R. (1981) *British Intelligence in the Second World War: Its Influence on Strategy and Operations*, vol. 2, London: Her Majesty's Stationery Office.

Hohenadel, K. (2004) "War told in depth . . . and in French," *Los Angeles Times*, 23 May, E25.

Holt, T. (1994) "The deceivers," *Military History Quarterly*, 7/4, 48–57.

Hoyt, E. (1981) *Guerrilla: Colonel von Lettow Vorbeck and Germany's East African Empire*, New York: MacMillan.

Hurley, N. (1976) "Using motion pictures to aid inter-cultural communication" in McLaurin, R., Rosenthal, C. and Shillings, S. (eds) *The Art and Science of Psychological Operations: Case Studies in Military Application*, vol. 2, Washington, DC: American Institutes for Research, 919–28.

Ikle, F. (1989) "The modern context" in Lord, C. and Barnett, F. (eds) *Political Warfare and Psychological Operations: Rethinking the U.S. Approach*, Washington, DC: National Defense University, 3–11.

Jaffe, J. (2002) "Dubya, willya turn the book over?" Wired.com, Online at www.wired.com/news/politics/0,1283,56430,00.html (accessed 21 August 2003).

Janis, I. (1972) *Groupthink: Psychological Studies of Policy Decisions and Fiascoes*, New York: Houghton Mifflin.

——— (1976) "Social Science Research, and PSYOP, Introduction Persuasion" in McLaurin, R., Rosenthal, C. and Shillings, S. (eds) *The Art and Science of Psychological Operations: Case Studies of Military Application*, vol. 2, Washington, DC: American Institutes for Research, 609–24.

Janowitz, M. (1958a) "The emancipation proclamation as an instrument of psychological warfare" in Daugherty, W. and Janowitz, M. (eds) *A Psychological Warfare Casebook*, Baltimore: Johns Hopkins University Press, 73–9.

——— (1958b) "Language idiom and accent in psychological warfare" in Daugherty, W. and Janowitz, M. (eds) *A Psychological Warfare Casebook*, Baltimore: Johns Hopkins University Press, 609–11.

Janowitz, M. and Marvick, E. (1958a) "US propaganda efforts and the 1948 Italian elections" in Daugherty, W. and Janowitz, M. (eds) *A Psychological Warfare Casebook*, Baltimore: Johns Hopkins University Press, 320–6.

——— (1958b) "The Katyn Incident," in Daugherty, W. and Janowitz, M. (eds) *A Psychological Warfare Casebook*, Baltimore: Johns Hopkins University Press, 352–6.

Jervis, R. (1976) *Perception and Misperception in International Politics*, Princeton, NJ: Princeton University Press.

Junger, S. (2002) *Fire*, New York: W.W. Norton.

JUSPAO Planning Staff (1976) "The Viet Cong Slogan Slip" in McLaurin, R., Rosenthal, C. and Shillings, S. (eds) *The Art and Science of Psychological Operations: Case Studies of Military Applications*, vol. 2, Washington, DC: American Institutes for Research, 682–4.

Katz, P. (1976) "A survey of PSYOP intelligence" in McLaurin, R., Rosenthal, C. and Shillings, S. (eds) *The Art and Science of Psychological Operations: Case Studies of Military Applications*, vol. 2, Washington, DC: American Institutes for Research, 478–94.

Keegan, J. (2003) *Intelligence in War: Knowledge of the Enemy from Napoleon to Al-Qaeda*, New York: Knopf.

Khong, Y. (1992) *Analogies at War: Korea, Munich, Dien Bien Phu, and the Vietnam Decisions of 1965*, Princeton, NJ: Princeton University Press.

King, D. (1997) *The Commissar Vanishes: The Falsification of Photographs and Art in Stalin's Russia*, New York: Metropolitan Books.

King, S. (2004) "It's not smoke and mirrors," *Los Angeles Times*, 26 September, E10.

Kinsley, M. (2005) "As through a glass darkly," *Los Angeles Times*, 6 March, M5.

Klein, J. (2004a) "The secrets of Reagan's success," *Time*, 14 June, 21.

——— (2004b) "What the swifties cost us," *Time*, 6 September, 21.

Knapp, S. (2004) "Seeing, not believing," e4 engineering.com, Online, available at http://www.e4engineering.com/story.aspx?uid = 9161fccf-1085–47f9–915a-e129d4c550ed&cuid = b96dad81–0ef4–4fcc-9e3d-a7bd9b6a4258 (accessed 29 July 2004).

Knightley, P. (1975) *The First Casualty: From the Crimea to Vietnam, the War Correspondents as Hero, Propagandist, and Myth Maker*, New York: Harcourt, Brace Jovanovich.

Kraul, C. (2004) "For good press, slip 'em pesos," *Los Angeles Times*, 30 August, A1, A4.

Kumata, H. and Schramm, W. (1958) "Propaganda theory of the German Nazis," in Daugherty, W. and Janowitz, M. (eds) *A Psychological Warfare Casebook*, Baltimore: Johns Hopkins University Press, 47–54.

Lansdale, E. (1976a) "Special operations against the Vietminh" in McLaurin, R., Rosenthal, C. and Shillings, S. (eds) *The Art and Science of Psychological Operations: Case Studies of Military Application*, vol. 2, Washington, DC: American Institutes for Research, 762–3.

——— (1976b) "Practical jokes" in McLaurin, R., Rosenthal, C. and Shillings, S. (eds) *The Art and Science of Psychological Operations: Case Studies of Military Application*, vol. 2, Washington, DC: American Institutes for Research, 767–70.

Latimer, J. (2001) *Deception in War: The Art of the Bluff, the Value of Deceit, and the Most Thrilling Episodes of Cunning in Military History, from the Trojan Horse to the Gulf War*, Woodstock, NY: Overlook Press.

Lemonick, M. (2004) "Chicken little alert," *Time*, 8 March, 69.
Lester, P. (2003) *Visual Communication: Images with Messages*, 3rd edn, Belmont, CA: Wadsworth Publishing.
Libicki, M. (2000) "What is information warfare?" in Gongora, T. and Riekhoff, H. (eds) *Toward a Revolution in Military Affairs? Defense and Security at the Dawn of the Twenty-First Century*, Westport, CT: Greenwood, 37–60.
Lieven, A. and Chambers, D. (2006) "The limits of propaganda," *Los Angeles Times*, 13 February, B13.
Linebarger, P. (1982) "The problem of the unintended audience" McLaurin, R. (ed.) *Military Propaganda, Psychological Warfare and Operations*, New York: Praeger, 339–41.
Los Angeles Times (2003) "Editor's note," *Los Angeles Times*, 2 April, A1, A6.
——— (2004) "Early warnings missed: a prison-abuse timeline," *Los Angeles Times*, 16 May, M2.
Luttwak, E. (1989) "Comment" in Lord, C. and Barnett, F. (eds) *Political Warfare and Psychological Operations: Rethinking the U.S. Approach*, Washington, DC: National Defense University, 163–7.
Macdonald, S. (2000) *Rolling the Iron Dice: Historical Analogies and Decisions to Use Military Force in Regional Contingencies*, Westport, CT: Greenwood.
——— (2002) "Hitler's shadow: historical analogies and the Iraqi invasion of Kuwait," *Journal of Diplomacy and Statecraft*, 13/4, 29–59.
McDonnell, P. (2004) "Fight for Fallouja began with a ruse," *Los Angeles Times*, 16 November, A9.
McGregor, D. and White R. (1990) *Reception and Response: Hearer Creativity and the Analysis of Spoken and Written Texts*, London: Routledge.
McLaurin, R., Rosenthal, C. and Shillings, S. (1976a) "Foreign policy and communications during the Hungarian uprising" in McLaurin R., Rosenthal, C. and Shillings, S. (eds) *The Art and Science of Psychological Operations: Case Studies of Military Application*, vol. 2, Washington, DC: American Institutes for Research, 382–5.
——— (1976b) "Tactical psyop" in McLaurin R., Rosenthal, C. and Shillings, S. (eds) *The Art and Science of Psychological Operations: Case Studies of Military Application*, vol. 2, Washington, DC: American Institutes for Research, 400–2.
——— (1976c) "Mainland pounded by balloons in psychological warfare," in McLaurin R., Rosenthal, C. and Shillings, S. (eds) *The Art and Science of Psychological Operations: Case Studies of Military Application*, vol. 2, Washington, DC: American Institutes for Research, 727–8.
Magnier, M. (2003) "Spy ship exhibit stirs up Japanese," *Los Angeles Times*, 15 June, A4.
——— (2005) "They can't handle the truth," *Los Angeles Times*, 28 February, A1, A6.
Margulies, L. (2004) "History channel is proven wrong," *Los Angeles Times*, 3 April, E2.
Massie, R. (2003) *Castles of Steel: Britain, Germany, and the Winning of the Great War at Sea*, New York: Random House.
Mazzetti, M. (2004) "PR meets psy-ops in war on terror," *Los Angeles Times*, 1 December, A1, A31.
——— (2006) "Military will keep planting articles in Iraq," *Los Angeles Times*, 4 March, A3.
Mazzetti, M. and Daragahi, B. (2005) "U.S. military covertly pays to run stories in Iraqi Press," *Los Angeles Times*, 30 November, A1, A12.
Mendez, A. (2000) *The Master of Disguise: My Secret Life in the CIA*, New York: Perennial.

Mendez, J. and Mendez, A. (2002) *Spy Dust: Two Masters of Disguise Reveal the Tools and Operations that Helped Win the Cold War*, New York: Atria Books.

Mercado, S. (2005) "Sailing the sea of OSINT in the Information Age," *Studies in Intelligence*, 48/3, 45–55.

Meyer, J. (2006) "CIA expands use of Drones in Terror War," *Los Angeles Times*, 29 January, A1, A28.

Miller, G. and Drogin, B. (2005) "Intelligence analysts whiffed on a 'curveball'," *Los Angeles Times*, 1 April, A1, A13.

Mock, J. and Larson, C. (1939) *Words that Won the War: The Story of the Committee of Public Information 1917–1919*, Princeton, NJ: Princeton University Press.

Mohler, J. and Livingston, G. (1982) "The Chinese Army and psychological warfare: past, present, promise" in McLaurin, R. (ed.) *Military Propaganda, Psychological Warfare and Operations*, New York: Praeger, 366–73.

Molander R., Riddile, A. and Wilson, P. (1996) *Strategic Information Warfare: A New Face of War*, Santa Monica, CA: RAND.

Montagu, E. (1969) *The Man Who Never Was*, New York: Bantam.

Morain, D. (2005) "Aides concede more mock news videos," *Los Angeles Times*, 10 March, B1, B11.

Morin, M. (2005) "Alleged kidnapping of soldiers may be a hoax," *Los Angeles Times*, 2 February, A6.

Moynahan, B. (1976) "Our poison pen war against the Nazis" in McLaurin, R., Rosenthal, C. and Shillings, S. (eds) *The Art and Science of Psychological Operations: Case Studies of Military Application*, vol. 2, Washington, DC: American Institutes for Research, 763–5.

Mullen, W. (ed.) (2005) *Battlefield Deception Operations*, CALL Bulletin, Combined Arms Training Activity, Online, available at http:call.army.mil/products/newsltrs/3-88/batt.htm (accessed 1 July 2005).

Murphy, J. (2003) "U.S. propaganda pitch halted," CBSnews.com, Online, available at www.cbsnews.com/stories/2003/01/16/world/main536756.shtml (accessed 6 February 2006).

Murray, D. (1958) "Operation American Dollar" in Daugherty, W. and Janowitz, M. (eds) *A Psychological Warfare Casebook*, Baltimore: Johns Hopkins University Press, 417–18.

Mycio, M. (2004) "Account of Chernobyl trip takes web surfers for a ride," *Los Angeles Times*, 6 July, A3.

Myers, D. (1996) *Social Psychology*, New York: McGraw-Hill, 5th edn.

Naftali, T. (1999) "The limits of intelligence," *Los Angeles Times Book Review*, 24 October, 7.

New York Times (2004) "The Times and Iraq," *New York Times*, 26 May, A10.

Nickell, J. (1994) *Camera Clues: A Handbook for Photographic Investigation*, Lexington: University Press of Kentucky.

Orwell, G. (1983) *1984*, New York: Harcourt Brace.

An OWI Psychological Warfare Team (1958) "Personnel Problems of OWI Assam Psywar Team" in Daugherty, W. and Janowitz, M. (eds) *A Psychological Warfare Casebook*, Baltimore: Johns Hopkins University Press, 185–91.

Pappas, D. (1976) "PSYOP and unit officer prerequisites" in McLaurin, R., Rosenthal, C. and Shillings, S. (eds) *The Art and Science of Psychological Operations: Case Studies of Military Application*, vol. 2, Washington, DC: American Institutes for Research, 272–82.

Ponsonby, A. (1928) *Falsehood in War-Time: Propaganda Lies of the First World War*, New York: E.P. Dutton & Co.

Postman, N. (1988) "Critical thinking in the electronic era" in Govier, T. (ed.) *Selected Issues in Logic and Communication*, Belmont, CA: Wadsworth Publishing, 11–19.

Prange, G., Goldstein, D. and Dillon, K. (1982) *Miracle at Midway*, New York: Penguin.

Preston, H., Monroe, J. and Raffa, A. (1976) "U.S. vulnerabilities as portrayed in the East German Television Film, 'Pilots in Pajamas' " in McLaurin, R., Rosenthal, C. and Shillings, S. (eds) *The Art and Science of Psychological Operations: Case Studies of Military Application*, vol. 2, Washington, DC: American Institutes for Research, 855–62.

Quang, P. (1976) "Better understanding of the role of newsreels and documentaries in the new situation" in McLaurin, R., Rosenthal, C. and Shillings, S. (eds) *The Art and Science of Psychological Operations: Case Studies of Military Application*, vol. 2, Washington, DC: American Institutes for Research, 1115–19.

Radu, M. (1990) "Contemporary insurgent political and psychological warfare" in Radvanyi, J. (ed.) *Psychological Operations and Political Warfare in Long-Term Strategic Planning*, New York: Praeger, 118–43.

Rainey, J. (2006) "More news outlets, fewer stories: new media 'paradox', " *Los Angeles Times*, 13 March, A7.

Rakowska-Harmstone, T. (1990) "Communist regimes' psychological warfare against their societies: the case of Poland" in Radvanyi, J. (ed.) *Psychological Operations and Political Warfare in Long-term Strategic Planning*, New York: Praeger, 92–112.

Rasmussen, C. (2003) "Muckraker's own life as compelling as his writing," *Los Angeles Times*, 11 May, B4.

Raviv, D. and Melman, Y. (1990) *Every Spy a Prince: The Complete History of Israel's Intelligence Community*, Boston, MA: Houghton Mifflin.

Reuters (1987) "Commando raids aimed to confuse Libyans," *Vancouver Sun*, 17 September, A5.

——— (2004) "Bush makes three mistakes while trying to cite one," Online, available at www.hypocrites.com/article17105.html (accessed 20 May 2004).

Richter, P. (2004) "U.S. struggles in war of ideas, panel says," *Los Angeles Times*, 25 July, A1, A14.

Riley, J. and Cottrell, L. (1958) "Research for psychological warfare" in Daugherty, W. and Janowitz, M. (eds) *A Psychological Warfare Casebook*, Baltimore: Johns Hopkins University Press, 536–45.

Rohatyn, D. (1988) "Propaganda talk" in Govier, T. (ed.) *Selected Issues in Logic and Communication*, Belmont, CA: Wadsworth Publishing, 73–92.

Romerstein, H. (1990) "Soviet active measures and propaganda: 'new thinking' and influence activities in the Gorbachev era" in Radvanyi, J. (ed.) *Psychological Operations and Political Warfare in Long-Term Strategic Planning*, New York: Praeger, 36–57.

Rosen, L. (2004) "Ad spoofs that are 'dynamite', " *Los Angeles Times*, 7 August, E8.

Rosenberg, H. (1997) "Exploring dubious traditions of news," *Los Angeles Times*, 7 March, F1.

——— (2003) "Seeking symbolic moments in the tides of history," *Los Angeles Times*, 10 April, A16.

Rubin, A. (2005) "To get the story, you have to get *to* the story," *Los Angeles Times*, 23 January, A10.

Rutten, T. (2004a) "Now smear this: a web of deceit," *Los Angeles Times*, 18 February, E1, E14.

——— (2004b) "N.Y. Times' latest misstep is also greatest," *Los Angeles Times*, 29 May, E1, E13.

——— (2005) "News' fate in today's corporate culture," *Los Angeles Times*, 22 January, E1, E6.
——— (2006) "When writing truth is a crime," *Los Angeles Times*, 25 February, E1, E16.
Sahagun, L. (2004) "Army trainers to become fighters in Iraq," *Los Angeles Times*, 17 October, B1.
Sanders, E. and Chu, H. (2004) "Cracks may be forming in cleric's rebel camp," *Los Angeles Times*, 21 August, A1, A14.
Scheer, R. (2003) "A soldier's dignity and humility," *Los Angeles Times*, 12 November, E1.
Schickel, R. (2005) "A blockbuster effect on art," *Los Angeles Times Book Review*, 13 March, R2.
Severin, W. and Tankard, J. (2001) *Communication Theories: Origins, Methods and Uses in the Mass Media*, New York: Longman.
Shaw, D. (2004) "Character assassination sullies the History Channel," *Los Angeles Times*, 7 March, E14.
Shultz, R. (1989) "Political strategies for revolutionary war" in Lord, C. and Barnett, F. (eds) *Political Warfare and Psychological Operations: Rethinking the U.S. Approach*, Washington, DC: National Defense University, 111–38.
Silverstein, K. (2004) "Oil adds sheen to Kazakh Regime," *Los Angeles Times*, 12 May, A1, A14–16.
——— (2005) "Democracy, or an exercise in fraud?" *Los Angeles Times*, 14 February, A1.
Sisson, E. (1958) "Sisson's account of Wilson's fourteen point speech" in Daugherty, W. and Janowitz, M. (eds) *A Psychological Warfare Casebook*, Baltimore: Johns Hopkins University Press, 89–96.
Smith, G. (2005) "Q Ships" Online, available at www.naval-history.net/WW1Navy BritishQShips.htm (accessed 3 March 2006).
Smith, L. (2004) "Web amplifies message of primitive executions," *Los Angeles Times*, June 30, A1, A12.
——— (2005) "New view of FDR includes disability," *Los Angeles Times*, 23 April, A1, A24–5.
Smith, P. (1989) "Comment" in Lord, C. and Barnett, F. (eds) *Political Warfare and Psychological Operations: Rethinking the U.S. Approach*, Washington, DC: National Defense University, 38–40.
Soller, M. (2004) "Be all that you can be—virtually," *Los Angeles Times*, 11 July, E4.
Sontag, S. (2003) "On photography (the short course)," *Los Angeles Times Book Review*, 27 July, R16.
Spurr, R. (1988) *Enter the Dragon: China's Undeclared War against the U.S. in Korea, 1950–53*, New York: New Market Press.
Stacy, M. (2005) "Lawmaker says photo altered," *The Bellingham Herald*, 4 August, A5.
Steed, H. W. (1958) "The Balfour Declaration of 1917" in Daugherty, W. and Janowitz, M. (eds) *A Psychological Warfare Casebook*, Baltimore: Johns Hopkins University Press, 262–3.
Summers, H. (1984) *On Strategy: A Critical Analysis of the Vietnam War*, New York: Dell.
Suvorov, V. (1984) *Inside the Soviet Army*, London: Granada.
Tayacan (1985) *Psychological Operations in Guerilla Warfare*, New York: Vintage.
Tesser, A. (1995) *Advanced Social Psychology*, New York: McGraw-Hill.
Thomas, K. (2004) "A tragic command performance," *The Los Angeles Times*, 26 February, E13.
Thompson, J. (1985) *No Picnic*, Glasgow: Fontana/Collins.

Turan, K. (2005) "How the U.S. waged peace after WWII," *Los Angeles Times*, 6 April, E1, E16.

——— (2004) "Horrific twists, choices of war," *Los Angeles Times*, 16 April, E6.

Uhlig, F. (1994) *How Navies Fight*, Annapolis: Naval Institute Press.

Underwood, M. (2003) "Reception analysis," Online, available at http://www.cultsock.ndirect.co.uk/MUHome/cshtml/index.html (accessed 7 February 2006).

USA Today (2004) "Why was pattern of abuse ignored for so long?" *USA Today*, 7 May, 11A.

Van Dam, N. (1982) "Israeli sectarian propaganda during the October 1973 War" in McLaurin, R. (ed.) *Military Propaganda, Psychological Warfare and Operations*, New York: Praeger, 356–65.

Vatis, M. (2001) "Government perspectives" in Alexander, Y. and Swetnam, M. (eds) *Cyber Terrorism and Information Warfare: Threats and Responses*, Ardsley, NY: Transnational Publishers, 3–12.

Vertzberger, Y. (1990) *The World in their Minds: Information Processing, Cognition, and Perception in Foreign Policy Decisionmaking*, Stanford: Stanford University Press.

Vieth, W. and Mazzetti, M. (2005) "Bush's chat with troops draws flak," *Los Angeles Times*, 14 October, A14.

Waltz, E. (1998) *Information Warfare: Principles and Operations*. Boston: Artech House.

Washington Post (2005) "Edward von Kloberg III, 63; Did PR for Hussein, other dictators," *Los Angeles Times*, 4 May, B12.

Weingarten, M. (2004) "War's new fronts," *Los Angeles Times*, 11 July, E4.

Westwood, J. (1984) *The History of the Middle East Wars*, Toronto: Bison Books.

Wheatley, G. and Hayes, R. (1996) *Information Warfare and Deterrence*, Washington, DC: National Defense University Press.

White, R. (1958) "Resistance to international propaganda," in Daugherty, W. and Janowitz, M. (eds) *A Psychological Warfare Casebook*, Baltimore: Johns Hopkins University Press, 617–25.

Williams, C. and Paddock, R. (2004) "Hostage in Iraq slain by captors," *Los Angeles Times*, 14 July, A1–A5.

Wood, H. (1972) *Vimy!* London: Corgi.

Zagorin, A. and Shannon, E. (2004) "Suspect snapshots," *Time*, 12 July, 5.

Zucchino, D. (2004) "Army stage-managed fall of Hussein statue," *Los Angeles Times*, 3 July, A28.

INDEX

For Product Safety Concerns and Information please contact our EU representative GPSR@taylorandfrancis.com
Taylor & Francis Verlag GmbH, Kaufingerstraße 24, 80331 München, Germany

www.ingramcontent.com/pod-product-compliance
Ingram Content Group UK Ltd.
Pitfield, Milton Keynes, MK11 3LW, UK
UKHW022129190926
475702UK00013B/1708
* 9 7 8 0 4 1 5 5 4 5 0 0 6 *